Advancing Professional Development through CPE in Public Health

Global Science Education

Professor Ali Eftekhari

Series Editor

Learning about the scientific education systems in the global context is of utmost importance now for two reasons. Firstly, the academic community is now international. It is no longer limited to top universities, as the mobility of staff and students is very common even in remote places. Secondly, education systems need to continually evolve in order to cope with the market demand. Contrary to the past when the pioneering countries were the most innovative ones, now emerging economies are more eager to push the boundaries of innovative education. Here, an overall picture of the whole field is provided. Moreover, the entire collection is indeed an encyclopaedia of science education and can be used as a resource for global education.

Series List:

The Whys of a Scientific Life
John R. Helliwell

Advancing Professional Development through CPE in Public Health
Ira Nurmala and Yashwant V. Pathak

Advancing Professional Development through CPE in Public Health

Ira Nurmala
Faculty of Public Health, Universitas Airlangga,
Surabaya, Indonesia

Yashwant V. Pathak
College of Pharmacy, University of South Florida,
Tampa, Florida USA
Adjunct Professor, Faculty of Public Health,
Universitas Airlangga, Surabaya, Indonesia

CRC Press
Taylor & Francis Group
Boca Raton London New York

CRC Press is an imprint of the
Taylor & Francis Group, an **informa** business

CRC Press
Taylor & Francis Group
6000 Broken Sound Parkway NW, Suite 300
Boca Raton, FL 33487-2742

© 2020 by Taylor & Francis Group, LLC
CRC Press is an imprint of Taylor & Francis Group, an Informa business

No claim to original U.S. Government works

Printed on acid-free paper

International Standard Book Number-13: 978-0-367-23636-6 (Hardback)

Visit the Taylor & Francis Web site at
http://www.taylorandfrancis.com

and the CRC Press Web site at
http://www.crcpress.com

Contents

List of Figures
and Tables

Series Preface

CONTRARY TO THE COMMON perception, the concept of education is not straightforward; both its purpose and its methods are subject to controversies. Plato was among the first who attempted to articulate the foundation of education by putting an emphasis on disciplines, which can directly help us to understand the universe.

Science education should be revisited in the changing climate and emerging needs of today. Both the concept and methods of education have been considerably evolved in the digital era. In the highly competitive market of higher education, higher-education institutions (HEIs) are too conservative to implement innovative changes. This series attempts to provide practical perspectives on various aspects of science education. In this book, the authors take us on a journey to the fundamental roots of science. With a long and fruitful experience in academia, they tell us where we are, why we are here, and how to survive.

The world is now more connected, and we can learn from a number of different systems, although direct interactions might be challenging due to social and cultural barriers. This series aims to provide a medium for a broad range of audiences to close this gap. I might add that the authors of this series do not necessarily share my opinions or concerns, of course.

The inception of this book series harks back to a friendly conversation with Hilary LaFoe, senior acquisitions editor at CRC Press. I should give at least half of the credit to her because the initial idea was hers. I would like to thank her because this project could not have materialised without her sincere commitment. I also encourage readers to give us feedback in order to help us take a small step towards furthering the concept of education at all levels according to the changing climate of globalisation.

Dr. Ali Eftekhari
Global Science Education

Acknowledgments

WE WOULD LIKE TO thank **Crystal Garcia**, College of Pharmacy, University of South Florida, who contributed to Chapter 2; **Khushali Vashi**, College of Public Health, University of South Florida, who contributed to Chapter 2; and **Rashmi Pathak**, College of Public Health, University of South Florida, who contributed to Chapters 1 and 2.

Authors

Ira Nurmala has more than 19 years of academic experience as a lecturer including the time she received her education in the public health field during her bachelor's, master's and PhD programs in Indonesia, the Netherlands and the U.S. She received a Fulbright Scholarship from the U.S. for her PhD and a scholarship from the Netherlands for her master's. She is currently Vice Dean of Research, Publication and Partnership in the Faculty of Public Health, Universitas Airlangga, Indonesia. Dr. Nurmala is a scholar, researcher and lecturer in public health majoring in health promotion with a special interest in continuing education for professionals, and youth health and a peer educator program.

Yashwant V. Pathak has more than 11 years of versatile administrative experience in an academic setting as dean, and more than 17 years as faculty and researcher in higher education after earning his PhD. He has worked in student admission, academic affairs, research, and graduate programs. He served as chair of the International Working Group for University of South Florida (USF) Health, including the Colleges of Medicine, Nursing, Public Health, Pharmacy and Physiotherapy, director of USF Center for Research and Education in Nanobioengineering, and now holds the position of Associate Dean for Faculty Affairs, College of Pharmacy. Dr. Pathak is an internationally recognized scholar, researcher and educator in the areas of health care education, nanotechnology, drug delivery systems and nutraceuticals.

CHAPTER 1

Public Health Education in the United States

1.1 INTRODUCTION: PUBLIC HEALTH EDUCATION IN THE UNITED STATES

The public health field has been recognized in the United States since the 1900s. In the period of 1900–1945, public health had a narrower interest than today (Wilner, 1973). During the 1900s, public health was interested in measuring communicable diseases, educating the community about selected illnesses, handling food and water, recording vital statistics, and treating diseases (Wilner, 1973). According to the American Public Health Association (APHA), public health is "the practice of preventing disease and promoting good health within groups of people, from small communities to entire countries." Public health professionals come from a variety of backgrounds and work in a variety of settings with the common goal of promoting population health (Gebbie, Rosenstock, and Hernandez, 2003a). The APHA also stated that public health professionals come from many educational backgrounds with the common purpose of protecting the health of a population by relying on policy and research. The core functions of public health are assessment, policy development and assurance. These functions act as a framework for public health operation. According to the Centers for Disease Control and Prevention, ten core essential services of public health are as follows: evaluate, monitor, diagnose, treat, inform, educate and empower, mobilize community partnerships, develop policies, enforce laws and ensure a competent workforce (Carlson, Chilton, Carso, and Beitsch, 2015).

Public health practice comprises individuals from a wide variety of backgrounds and a wide variety of work settings working together toward one common goal to promote population health. "A public health professional is a person educated in public health or a related discipline who is employed to improve health through a population focus" (Gebbie et al., 2003a, p. 30). Public health practitioners must ensure the quality of service given to the community. This means that public health is everybody's concern. Health and illness are a part of life, but people should be educated about preventing some health problems especially when more people are exposed to a polluted environment or live a certain lifestyle that can cause health problems to themselves or to their community.

In the past decade, many efforts have been conducted toward improving the health of individuals and communities through health education efforts. Health education efforts are not only conducted in schools, but also in public spaces, such as grocery stores and public transportation, among other places. These efforts may vary depending on the urgency of the messages that need to be delivered in these public places. A reminder to wash one's hands in a public restroom is one of simplest health education efforts that have been conducted to raise public awareness of personal hygiene. One of the prominent professions in the public health field that concerns educating individuals and communities to promote health and prevent diseases is called "public health educators (PHEs)." Public health educators are drawn from a diverse range of disciplines and backgrounds and may or may not have formal qualifications in the field (i.e., no professional preparation or post-graduate qualifications in health education). A public health educator is defined as "a professionally prepared individual who serves in a variety of roles and is specifically trained to use appropriate educational strategies and methods to facilitate the development of policies, procedures, interventions, and systems conducive to the health of individuals, groups, and communities" (Joint Committee on Health Education and Promotion Terminology, 2002, p. 6). According to this definition, public health educators serve the community through education efforts to improve the community's health status.

Currently, at least 250 academic programs exist in colleges and universities that aim to educate professional PHEs. More than 12,000 professionals from these colleges and universities have received the designation of Certified Health Education Specialist (CHES) nationwide (National Commission for Health Education Credentialing, 2006). However, study shows most public health professionals do not engage in activities that can enhance their professional development. The public health workforce is not yet prepared to meet the challenge to ensure the quality of practice in the context of rapid social change (Allegrante, Moon, Auld, and Gebbie, 2001). In addition, 21.2% of public health professionals are up to date on their work of implementing programs, and 60% did not conduct research or professional development activities (Johnson, H. H., Glascoff, Lovelace, Bibeau, and Tyler, 2005). Moreover, there is a gap between the current skills and the skills needed in their professional work.

1.2 RESPONSIBILITIES AND COMPETENCIES FOR PUBLIC HEALTH EDUCATORS

To pursue a profession in the field of health education and promotion, more than just acquiring credentials is required. The core responsibilities, competencies, and sub-competencies provide a comprehensive description of the profession, illustrating the skills necessary to perform the daily tasks as a public health educator or health education specialist.

In the following documents, you can access the "Seven Areas of Responsibility" for health education specialists determined by the latest job analysis studies, which go beyond credentials and into the heart of the profession, including competencies such as planning and evaluation, administration, communication, promotion and more.

The Seven Areas of Responsibility present the required skills and expertise needed for a position in the field of health education and promotion. The Seven Areas of Responsibility were verified by the 2015 Health Education Specialist Practice Analysis (HESPA) project and serve as the basis of both the CHES® and the MCHES® exams. The HESPA study led to an adjustment to the health education model to include the term "health promotion." The rationale for the terminology change to "health education/health promotion" was that it adds clarity to the scope of the health education specialist's role both within and external to the profession and would more comprehensively describe the profession (National Commission for Health Education Credentialing Inc., 2013). As professionals, public health educators are also bound by responsibilities that were established since 1985. In 2004, the National Health Educator Competencies Update Project (CUP) Model, which was funded by the American Association for Health Education (AAHE), the National Commission for Health Education Credentialing (NCHEC), and the Society for Public Health Education (SOPHE) revised the Seven Areas of Responsibility for public health educators (National Commission for Health Education Credentialing Inc., 2013). The Competencies Update Project that was conducted from 1998 to 2004 addressed what health educators do in practice, the degree to which the definition of the entry-level health educator role remains up-to-date; and the validation of advanced-level competencies (Gilmore, Olsen, Taub, and Connell, 2005). Each of these Seven Areas of Responsibility has its own competencies (National Commission for Health Education Credentialing Inc., 2013) (Figure 1.1).

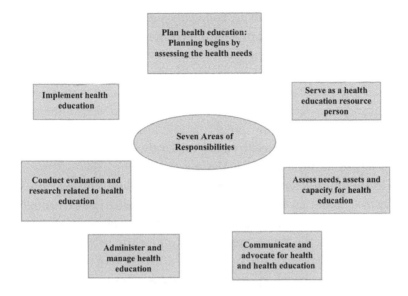

FIGURE 1.1   Seven areas of responsibility of public health workers.

1.3 AREAS OF RESPONSIBILITY

1.3.1 Area of Responsibility I: Assess Needs, Assets and Capacity for Health Education

All Seven Areas of Responsibility give a general idea of what health education specialists do without having to provide the details that are necessary to practice health education. One major area of responsibility for health education specialists is the first area of responsibility, which is to assess needs, assets, and the capacity for health education. A need assessment is a systematic process that helps health education specialists determine health problems of the priority population, available assets within that population and their overall capacity to address that particular health issue. The first step in assessment is to identify existing resources, research designs, methods and instruments that are relevant to applying the assessment process.

To successfully conduct the assessment, health educators will need to identify health-related databases and valid sources of data with appropriate instruments, and apply survey techniques such as quantitative or qualitative methods in collecting health information. Then, health educators will need to examine factors that influence health behaviors and gaps in health education programs that may hinder the priority population's learning process. If there are no hindrances, program planners will continue to determine the available health education programs and analyze the capacity for developing need-based health education. Lastly, health educators need to prioritize health education in the sequence in which they will teach.

In the international work setting, health education specialists may have to stay up-to-date on the emerging health issues through surveys done in hospital and clinical settings, or go from door to door to ask the target population to join the surveys. Health problems that are identified may also be different than in the United States. Instead of dealing with heart disease, obesity and diabetes, health education specialists in these settings may be dealing more with issues such as starvation, malnutrition and parasitic and bacterial infections in a larger population. In addition, health education specialists need to gain support from stakeholders such as administrators, government officials or businesses in the country. When the support of the stakeholders is being obtained, health education specialists will have to prioritize what health education is relevant to the target population.

1.3.2 Area of Responsibility II: Plan Health Education: Planning Begins by Assessing Health Needs

Addressing the problems and concerns of the population is the secondary area which should be addressed. Once the need assessment has been determined, the program planning process can be set in motion. It is important to recruit stakeholders or other administrators early on in program planning, so that they can help develop the program. To be effective, the health educators should have strong written and oral communications skills, leadership ability and expertise to help the stakeholders reach consensus on the issues of interest. After getting stakeholder involvement, the health education specialists need to establish goals and objectives specific to the proposed health education program. Next, health education specialists need to design theory-based strategies and interventions to achieve state objectives. Relevant educational methods should be selected as well as locating resources. Then, the health educators need to plan the scope and sequence to deliver health education and start evaluating if the program's implementation is working.

Planning health education in an international work setting may look a little bit different than in the United States. Health education specialists may have to identify all the different ways that they could distribute information to the priority population. In an international work setting, health educators may have to distribute information through local newspapers, radio, television or mobile communication systems, such as hooking up an old stereo system to a car and driving it through the community. As the car is in motion, it will play the health message aloud so that everyone in the community could hear it. This is why it is important that health educators have the support and interest of the administrators or the government officials because, with their support, the program will have all the resources it needs to be effective. Furthermore, the priority population will see the program as credible when it has been sponsored by a local organization or the government.

1.3.3 Area of Responsibility III: Implement Health Education

After the need assessment is conducted and health education is planned, it is time for the health educators to implement the health program. This area of the responsibility is probably the most enjoyable due to the fact that the health educators are finally getting some hands-on experience now that the program is put into action. To effectively implement the program, health education specialists must have a full understanding of the priority population. This will help health educators eliminate cultural and language barriers when implementing health education plans. After obtaining additional information about the priority population, health education specialists must write objectives that are suitable to the program and select media and methods that are relevant to the target audience. Health educators must also conduct the program as planned and modify the plan of action as needed.

Like the program planning phase, health education specialists must be creative in many ways to implement the health program. The lesson plans should be fun and educational to teach the target audience whether in classrooms or through television broadcasts. Health education specialists must identify health lessons that would be acceptable to the social norms of the community. To achieve this, health educators must know and understand the priority population in order to eliminate cultural and language disparities. This is critical, especially in an international work setting, because culture and custom is a very important factor. Furthermore, the priority population will be most likely to listen to health educators and change their behavior accordingly towards their health problems.

1.3.4 Area of Responsibility IV: Conduct Evaluation and Research Related to Health Education

All health education specialists are expected to have the ability to thoroughly review researched articles and apply the findings to program planning as well as conduct effective evaluations. Health education specialists must be able to conduct evaluation of policy, projects and programs. Before evaluation, the health education specialist must first have realistic and measurable objectives for the health program. After the objectives are in place, the health educators must develop a plan that will accurately assess if the program objectives have been met. Some of the instruments that health education specialists can use in the evaluation process are tests, surveys, behavior observation, tracking epidemiology data, etc. They must be analyzed and interpreted, which can be done through the use of descriptive statistics or qualitative methods. Next, the health education specialists must compare the results to evaluation or other research findings and propose possible explanations of findings. Lastly, before applying the findings from evaluation to program development, health educators must identify possible limitations of findings. Doing so will help maintain support and funding in a competitive environment.

Good relationships between health educators and their audience will help health education specialists fully understand the issue, because the target population will be more willing to share on a personal level. For this particular reason, health educators will be able to know if the program is really effective. Evaluation of health education programs can be gained through surveys, interviews or going from house to house and asking the target population if the health program is working. After the program evaluation is done, health education specialists put the findings from evaluation into policy analysis and program development.

1.3.5 Area of Responsibility V: Administer and Manage Health Education

Administration, management and coordination are needed in order to have a successful health program. Although some administrative tasks are performed by professionals at an advanced level, administration and management responsibilities are handled by experienced health education specialists who are at the entry level. Health education specialists often become program administrators or staff supervisors. Some of the qualities that a good manager and supervisor may have include effective interpersonal skills and leadership skills with managing fiscal resources, knowledge of budgeting, task assignments and performance evaluation. Health education specialists must obtain acceptance and support from stakeholders who are responsible for health education by explaining to them how program goals align with organizational structure, mission and goals. After gaining the stakeholders' support, health education specialists must identify other potential partners. These potential partners could be health education specialists themselves.

In this area of responsibility, health education specialists with CHES and minimum experience can hold a high-level administrative job because it is rare to have health education specialists with CHES working in the international work settings. Like in the United States work setting, health education specialists in an international work setting must also possess the skills of good communication. They must be able to communicate to the target population as well as to the potential stakeholders in order to get them on board with the health program. In addition, health education specialists must also have the skill of budgeting different resources such as money, and the knowledge of health information. This is especially important because money is scarce and health information is limited; thus, knowing how to prioritize will allow the program to become effective.

1.3.6 Area of Responsibility VI: Serve as a Health Education Resource Person

Health education specialists are often called to serve as resource people to the target population with valid and reliable health information and materials. Health education specialists must be aware of the various resources at the state and national levels and must have the skill to access information from those resources. Being computer literate may help many health educators access health information through the Internet, library databases or other national online databases. Health education specialists must be able to evaluate and select appropriate resource material to educate the intended audience. As part of this process, it is probably necessary for health educators to develop effective educational pamphlets or brochures for distributing information to the priority population. Lastly, being a resource person, health educators must establish advisory relationships in order to facilitate collaborative efforts to achieve program goals.

This responsibility is very critical and often needed in the international setting. Health information and people who are health literate are scarce in the Third World. There are Internet databases; however, there are very few people who are literate and could use the information and interpret it for the general population to understand. Health education specialists in this work setting must also be knowledgeable about health by staying current about emerging health problems. In addition, health education specialists must have good communication skills to facilitate an effective health program.

1.3.7 Area of Responsibility VII: Communicate and Advocate for Health and Health Education

In this area of responsibility, health education specialists are responsible for providing information to various groups of people, including other health education specialists, health professionals, consumers, students, employers and employees. Health education specialists must identify current and emerging issues that may influence health and health education, and access accurate resources related to the needs assessments. After health education specialists identify health issues among the intended population, they must develop a variety of communication strategies, methods and techniques of how to improve the health problems to distribute to the priority population. Some of the communications strategies that health education specialists can use are mass media communication, such as written or oral. In addition, health education specialists use their professional skills to interpret and filter difficult scientific concepts so that the priority population can understand the information that is needed to improve their health. Health education specialists are also responsible for advocating health and health education to the target population. By advocating, health education specialists initiate and support legislation, rules, policies and procedures that will enhance the population's health. Health education specialists are required to advocate for the health and health education profession as well. Health education specialists advocate for their profession by educating potential employers about the value of hiring professionally trained health education specialists with CHES and MCHES credentials.

Health education specialists have the responsibility of providing information about health to various populations. Just like in Area of Responsibility VI, the health education specialists must stay up-to-date with the current as well as old health issues in order to serve the target population in the international work setting effectively. They must be able to read difficult scientific concepts and interpret them in a language that is easy for the general population to understand. After interpreting difficult scientific concepts of health information, health educators must identify ways to distribute them to the various groups. Then they must act as health advocates for health as well as for their profession by training more people to become professional health educators.

All the Seven Areas of Responsibility, competencies and sub-competencies allow health education specialists to practice their profession effectively. These responsibilities, competencies and sub-competencies do not function independently from one another and are highly interrelated. Health education specialists are required to have excellent communications skills and skills to identify and gather appropriate resources. They must be able to conduct accurate need assessment, plan, implement and evaluate the education program. Health education specialists must also be able to act as the administrators of the program, as resource people, and as advocates for both health education and their profession. All Seven Areas of Responsibility are critical for effective health education to take place. This is why it is very important that health education specialists know and understand them so that they can effectively help the population in need.

1.4 COMPETENCY STATEMENTS FOR PUBLIC HEALTH WORKERS DEVELOPED IN THE UNITED STATES

Professional public health associations ensure that public health educators maintain their professional credibility by addressing the multiple needs or matters that public health educators experience in their professional work. Public health educators state that the main reasons why they become members of a professional organization are the ability to maintain CHES certification, advance in the profession and to network with other professionals (Thackeray, Neiger, and Roe, 2005). There are no dominant professional associations at the national level, even APHA, SOPHE, AAHE captured only 55% of the national market (Thackeray et al., 2005). In the United States, a national credentialing system, administered by NCHEC, was established in 1988 (Taub, Allegrante, Barry, and Sakagami, 2009). NCHEC establishes national standards for the practice of public health educators, administers a national certification examination, and regulates continuing education requirements that are designed to promote continued professional development for those members who are certified (Taub et al., 2009). The idea of advanced credentialing was first introduced more than 20 years ago (Dennis and Lysoby, 2010). The percentage of questions on the national certification exam is based on the results of the 2009 Health Education Job Analysis (HEJA), and they reflect the percentage of time spent in each of the competency areas by practicing health educators (Dennis and Lysoby, 2010). In 2010, a competency-based framework was constructed to describe background information for public health educators' professional development (Dennis and Lysoby, 2010). In the United States, a variety of accreditation processes are available to academic programs in colleges and universities to enhance the quality of professional preparation (Allegrante et al., 2001). One of the examples is the accreditation by an independent agency, such as Council on Education for Public Health (CEPH) that is recognized by the U.S. Department of Education to give accreditation to schools of public health and public health programs (University of Georgia, College of Public Health, 2013). Although many accreditation processes are voluntary, this process provides standards for the academic professional preparation of public health educators (Allegrante et al., 2001). The academic professional preparation for public health educators is to prepare the professional to work in the following areas of competency:

AREA I: ASSESS NEEDS, ASSETS, AND CAPACITY FOR HEALTH EDUCATION
Competency 1.1: Plan assessment process
Sub-competencies

1.1.1 Identify existing and needed resources to conduct assessments

1.1.2 Apply theories and models to develop assessment strategies

1.1.3 Develop plans for data collection, analysis, and interpretation

1.1.4 Integrate research designs, methods, and instruments into assessment plans

Competency 1.2: Access existing information and data related to health
Sub-competencies

1.2.1 Identify sources of data related to health

1.2.2 Critique sources of health information using theory and evidence from the literature

1.2.3 Select valid sources of information about health

1.2.4 Identify gaps in data using theories and assessment models

1.2.5 Establish collaborative relationships and agreements that facilitate access to data

1.2.6 Conduct searches of existing databases for specific health-related data

Competency 1.3: Collect quantitative and/or qualitative data related to health
Sub-competencies

1.3.1 Collect primary and/or secondary data

1.3.2 Integrate primary data with secondary data

1.3.3   Identify data collection instruments and methods

1.3.4   Develop data collection instruments and methods

1.3.5   Train personnel and stakeholders regarding data collection

1.3.6   Use data collection instruments and methods

1.3.7   Employ ethical standards when collecting data

Competency 1.4: Examine relationships among behavioral, environmental and genetic factors that enhance or compromise health
Sub-competencies

1.4.1   Identify factors that influence health behaviors

1.4.2   Analyze factors that influence health behaviors

1.4.3   Identify factors that enhance or compromise health

1.4.4   Analyze factors that enhance or compromise health

Competency 1.5: Examine factors that influence the learning process
Sub-competencies

1.5.1   Identify factors that foster or hinder the learning process

1.5.2   Identify factors that foster or hinder attitudes and beliefs

1.5.3   Analyze factors that foster or hinder attitudes and beliefs

Competency 1.6: Examine factors that enhance or compromise the process of health education
Sub-competencies

1.6.1   Determine the extent of available health education programs, interventions, and policies

1.6.2   Assess the quality of available health education programs, interventions, and policies

1.6.3 Identify existing and potential partners for the provision of health education

1.6.4 Assess social, environmental, and political conditions that may impact health education

1.6.5 Analyze the capacity for developing needed health education

1.6.6 Assess the need for resources to foster health education

Competency 1.7: Infer needs for health education based on assessment findings
Sub-competencies

1.7.1 Analyze assessment findings

1.7.2 Prioritize health education needs

1.7.3 Identify emerging health education needs

1.7.4 Report assessment findings

AREA II: PLAN HEALTH EDUCATION
Competency 2.1: Involve priority populations and other stakeholders in the planning process
Sub-competencies

2.1.1 Incorporate principles of community organization

2.1.2 Identify priority populations and other stakeholders

2.1.3 Communicate need for health education to priority populations and other stakeholders

2.1.4 Develop collaborative efforts among priority populations and other stakeholders

2.1.5 Elicit input from priority populations and other stakeholders

2.1.6 Assess resources needed to achieve objectives

Competency 2.2: Develop goals and objectives
Sub-competencies

2.2.1   Identify desired outcomes utilizing the needs assessment results

2.2.2   Assess resources needed to achieve objectives

Competency 2.3: Select design strategies and interventions
Sub-competencies

2.3.1   Design theory-based strategies and interventions to achieve stated objectives

2.3.2   Comply with legal and ethical principles in designing strategies and interventions

2.3.3   Apply principles of cultural competence in selecting and designing strategies and interventions

2.3.4   Pilot test strategies and interventions

Competency 2.4: Develop a scope and sequence for the delivery of health education
Sub-competencies

2.4.1   Determine the range of health education needed to achieve goals and objectives

2.4.2   Select resources required to implement health education

2.4.3   Use logic models to guide the planning process

2.4.4   Analyze the opportunity for integrating health education into other programs

2.4.5   Develop a process for integrating health education into other programs

Competency 2.5: Address factors that affect implementation
Sub-competencies

2.5.1   Identify factors that foster or hinder implementation

2.5.2 Analyze factors that foster or hinder implementation

2.5.3 Use findings of pilot to refine implementation plans as needed

2.5.4 Develop a conducive learning environment

AREA III: IMPLEMENT HEALTH EDUCATION
Competency 3.1: Implement a plan of action
Sub-competencies

3.1.1 Assess readiness for implementation

3.1.2 Collect baseline data

3.1.3 Use strategies to ensure cultural competence in implementing health education plans

3.1.4 Use a variety of strategies to deliver a plan of action

3.1.5 Promote plan of action

3.1.6 Apply theories and models of implementation

3.1.7 Launch plan of action

Competency 3.2: Monitor implementation of health education
Sub-competencies

3.2.1 Monitor progress in accordance with timeline

3.2.2 Assess progress in achieving objectives

3.2.3 Modify plan of action as needed

3.2.4 Monitor use of resources

3.2.5 Monitor compliance with legal and ethical principles

Competency 3.3: Train individuals involved in implementation of health
Sub-competencies

3.3.1 Select training participants needed for implementation

3.3.2   Demonstrate a wide range of training strategies

3.3.3   Deliver training

AREA IV: CONDUCT EVALUATION AND RESEARCH RELATED TO HEALTH EDUCATION
Competency 4.1: Develop plans for evaluation and research
Sub-competencies

4.1.1   Assess feasibility of conducting evaluation/research

4.1.2   Critique evaluation and research methods and findings in the related literature

4.1.3   Synthesize information found in the literature

4.1.4   Assess the merits and limitations of qualitative and quantitative data

4.1.5   Identify existing data collection instruments

4.1.6   Critique existing data collection instruments for evaluation

4.1.7   Develop data analysis plan for evaluation

4.1.8   Apply ethical standards in developing the evaluation/ research plan

Competency 4.2: Design instruments to collect evaluation/ research data
Sub-competencies

4.2.1   Identify useable questions from existing instruments

4.2.2   Write new items to be used in data collection for evaluation

4.2.3   Establish validity of data collection instruments

4.2.4   Establish reliability of data collection instruments

Competency 4.3: Collect and analyze evaluation/research data
Sub-competencies

4.3.1 Collect data based on the evaluation/research plan

4.3.2 Monitor data collection and management

4.3.3 Analyze data using descriptive statistics

4.3.4 Analyze data using inferential and/or other advanced statistical methods

4.3.5 Analyze data using qualitative methods

4.3.6 Apply ethical standards in collecting and analyzing data

Competency 4.4: Interpret results of the evaluation/research
Sub-competencies

4.4.1 Compare results to evaluation/research questions

4.4.2 Compare results to other findings

4.4.3 Propose possible explanations of findings

4.4.4 Identify possible limitations of findings

4.4.5 Develop recommendations based on results

Competency 4.5: Apply findings from evaluation/research
Sub-competencies

4.5.1 Communicate findings to stakeholders

4.5.2 Apply findings in policy analysis and program development

AREA V: ADMINISTER AND MANAGE HEALTH EDUCATION
Competency 5.1: Obtain acceptance and support for programs
Sub-competencies

5.1.1 Provide support for individuals who deliver professional development opportunities

5.1.2 Explain how program goals align with organizational structure, mission, and goals

Competency 5.2: Demonstrate leadership
Sub-competencies

5.2.1 Conduct strategic planning

5.2.2 Analyze an organization's culture in relationship to health education goals

5.2.3 Develop strategies to reinforce or change organizational culture to achieve health education goals

5.2.4 Comply with existing laws and regulations

5.2.5 Adhere to ethical standards of the profession

5.2.6 Facilitate efforts to achieve organizational mission

5.2.7 Analyze the need for a systems approach to change

5.2.8 Facilitate needed changes to organizational cultures

Competency 5.3: Manage human resources
Sub-competencies

5.3.1 Develop volunteer opportunities

5.3.2 Demonstrate leadership skills in managing human resources

5.3.3 Apply human resource policies consistent with relevant laws and regulations

5.3.4 Evaluate qualifications of staff and volunteers needed for programs

5.3.5 Recruit volunteers and staff

5.3.6 Apply appropriate methods for team development

5.3.7 Model professional practices and ethical behavior

5.3.8 Evaluate performance of staff and volunteers

Competency 5.4: Facilitate partnerships in support of health education
Sub-competencies

    5.4.1   Facilitate partner relationship (s)

AREA VI: SERVE AS A HEALTH EDUCATION RESOURCE PERSON
Competency 6.1: Obtain and disseminate health-related information
Sub-competencies

    6.1.1   Assess information needs

    6.1.2   Identify valid information resources

    6.1.3   Critique resource materials for accuracy, relevance, and timeliness

    6.1.4   Convey health-related information to priority populations

    6.1.5   Convey health-related information to key stakeholders

Competency 6.2: Provide training
Sub-competencies

    6.2.1   Identify priority populations

Competency 6.3: Serve as a health education consultant
Sub-competencies

    6.3.1   Assess needs for assistance

    6.3.2   Prioritize requests for assistance

    6.3.3   Define parameters of effective consultative relationship

    6.3.4   Establish consultative relationships

    6.3.5   Facilitate collaborative efforts to achieve program goals

    6.3.6   Apply ethical principles in consultative relationships

AREA VII: COMMUNICATE AND ADVOCATE FOR HEALTH AND HEALTH EDUCATION

Competency 7.1: Assess and prioritize health information and advocacy needs

Sub-competencies

7.1.1 Identify current and emerging issues that may influence health and health education

7.1.2 Access accurate resources related to identified issues

7.1.3 Analyze the impact of existing and proposed policies on health

7.1.4 Analyze factors that influence decision makers

Competency 7.2: Identify and develop a variety of communication strategies, methods, and techniques

Sub-competencies

7.2.1 Create messages using communication theories and models

7.2.2 Tailor messages to priority populations

7.2.3 Incorporate images to enhance messages

7.2.4 Select effective methods or channels for communicating to priority populations

7.2.5 Pilot test messages and delivery methods with priority populations

7.2.6 Revise messages based on pilot feedback

Competency 7.3: Promote the health education profession individually and collectively

Sub-competencies

7.3.1 Use techniques that empower individuals and communities to improve their health

7.3.2 Employ technology to communicate to priority populations

7.3.3 Evaluate the delivery of communication strategies, methods, and techniques

Competency 7.4: Engage in health education advocacy
Sub-competencies

7.4.1 Engage stakeholders in advocacy

7.4.2 Develop an advocacy plan in compliance with local, state, and/or federal policies and procedures

7.4.3 Comply with ˙ organizational policies related to participating in advocacy

7.4.4 Communicate the impact of health and health education on organizational and socio-ecological factors

7.4.5 Use data to support advocacy messages

7.4.6 Implement advocacy plans

7.4.7 Incorporate media and technology in advocacy

7.4.8 Participate in advocacy initiatives

Competency 7.5: Influence policy to promote health
Sub-competencies

7.5.1 Identify the significance and implications of health policy for individuals, groups, and communities

7.5.2 Advocate for health-related policies, regulations, laws, or rules

7.5.3 Employ policy and media advocacy techniques to influence decision makers

Competency 7.6: Promote the health education profession
Sub-competencies

7.6.1 Develop a personal plan for professional growth and service

7.6.2 Describe state-of-the-art health education practice

7.6.3 Explain the major responsibilities of the health education specialist in the practice of health education

7.6.4 Explain the role of health education associations in advancing the profession

7.6.5 Explain the benefits of participating in professional organizations

7.6.6 Facilitate professional growth of self and others

7.6.7 Explain the history of the health education profession and its current and future implications for professional practice

7.6.8 Explain the role of credentialing in the promotion of the health education profession

7.6.9 Engage in professional development activities

7.6.10 Serve as a mentor to others

7.6.11 Develop materials that contribute to the professional literature

7.6.12 Engage in service to advance the health education profession

1.5 PUBLIC HEALTH SYSTEM IN THE UNITED STATES FEDERAL GOVERNMENT

The federal government assumes an extensive job in the general well-being framework of the nation. It overviews the population's health care status and needs, sets strategies and measures, passes laws and directions. Further, it underpins biomedical and healthcare administrations, which helps conveying individual healthcare. The federal government does these tasks for the most part through two assigned forces: the ability to manage interstate trade and the ability spend for the general welfare. The government's administrative activities, for example, marking dangerous substances, lie in the ability to manage interstate business. Its administrative projects, for example, the cleanup of dangerous substances or financing individual health care administrations through Medicaid and Medicare programs, lie in its capacity to be in charge of and spend for the general welfare (Grad, 1981). At present, the principal federal unit in charge of the population's general well-being is the United States Public Health Service in the Department of Health and Human Services. The second main unit is the Health Care Financing Administration, which is also in the Department of Health and Human Services. Further, the federal government assists the states when they lack the resources or expertise to effectively work in health emergencies (Omenn, 1982). The president of the United States and his cabinet are responsible for the election of the secretary of the Department of Health and Human Services and the head of the Public Health Service, the assistant secretary of health. There are 55 state health offices in the nation (50 states in addition to the District of Columbia, Guam, Puerto Rico, American Samoa, and the U.S. Virgin Islands). Each state health office is coordinated by a well-being chief or secretary of health. Each likewise has a state health officer who is responsible for therapeutic specialist in the state. In numerous states, the state health officer is the chief health officer. State health officers are delegated either by the state representative, the State Board of Health, or an organization head (Council of State Governments, 1987) (Table 1.1).

Public health in the United States is regulated by the following organizations: (1) the National Institutes of Health; (2) the Centers for Disease Control and Prevention; (3) the U.S. Food and Drug Administration; (4) the Substance Abuse and Mental Health Services Administration; (5) the Health Resources and Services Administration; and (6) the Agency for Toxic Substances and Disease Registry. The National Institutes of Health mainly deals with research; it conducts and administers the research projects. The Centers for Disease Control and Prevention is the main epidemiologic unit of the nation; it directly deals with state and local departments

TABLE 1.1   State Health Officers

	Number of States (n = 49)
A. *Appointment Procedures*	
Appointed by Governor	33
Appointed by Agency Director	10
Appointed by State Board of Health	6
	Number of States (n = 46)
B. *Education and Experience Requirements*	
Medical Degree	25
Medical Degree + Master's Degree of Public Health	8
Medical Degree + Public Health Experience	10
Public Health Experience	3

Source: From Institute of Medicine, Division of Health Care Services, Committee for the Study of the Future of Public Health. 1988. The Future of Public Health. Washington, D.C.: National Academy Press. Retrieved from https://books.google.co.id/books?id=pHDt0C3LEag C&pg=PA173&lpg=PA173&dq=American+Medical+Association ,+Department+of+State+Legislation,+1984&source=bl&ots=Jh2 k24bxcP&sig=ACfU3U3lfQePyBBNLxND_0_UpRBIpY450g&hl=e n&sa=X&ved=2ahUKEwjLvqHej5jhAhXEfn0KHaD7DSoQ6AEw.

and provides them with technical assistance. The Health Resources and Services Administration is concerned with development of resources and health workforce. The Substance Abuse and Mental Health Administration's job is to oversee lessening the impact of substance abuse and mental illness on the American population. The U.S. Food and Drug Administration regulates the standards of food and drugs safety (Figure 1.2).

Sometimes these organizations directly provide services to specific groups of people; for instance, the Indian Health Services office in the Health Resources and Services Administration, through the help of its government, provides services to Native Americans and Eskimos (Hanlon and Pickett, 1984). Other working divisions of the Department of Health and Human Services are essentially geared toward human and social administrations. These divisions, despite the fact that they are

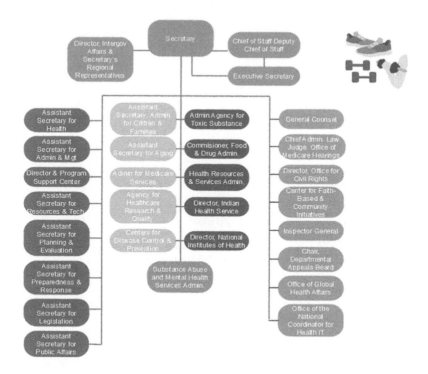

FIGURE 1.2 U.S. Department of Health and Human Services (HHS) organizational chart. (From https://viewer.edrawsoft.com/public/s/b1dce076075414.)

not assigned explicitly for well-being, direct numerous well-being-related activities. For instance, the Office of Human Development Services houses the Administration on Aging and the Administration on Developmental Disabilities, the two of which are engaged with long-term medical services matters. Other federal agencies with public health responsibilities are: the Department of Education, the Department of Transportation, the Department of Veterans Affairs, the Department of Homeland Security, the Department of Agriculture, the Department of Housing and the Environmental Protection Agency (Grad, 1981). In the United States, the individual states and the local governments play an important role in regulating public health and policies. The states are the key administrative element in charge of ensuring the general well-being of Americans. They lead a wide scope of activities in well-being. State offices gather and investigate data, direct examinations, plan, set arrangements and norms, carry national and state orders, oversee and manage ecological, instructive, and individual health administrations, and guarantee access to social insurance for underserved people; they are engaged with assets improvement; and they react to perils and emergencies (Hanlon and Pickett, 1984). States carry out a large portion of their duties through their police control; they have the power "to sanction and uphold laws to secure and advance the wellbeing, ethics, harmony, solace, and general welfare of the general population" (Grad, 1981). The General Assembly deals with the insurance, enhancement, and safeguarding of the public health as a fundamental right for the general welfare of the natives of the Commonwealth. The State Board of Health and the State Health Commissioner, in collaboration with the State Department of Health, will direct and give a thorough program enlisting preventive, remedial, therapeutic, and environmental health services, teach the citizenry in wellbeing and ecological issues, create and actualize wellbeing asset designs, gather and save imperative records and wellbeing insights, aid, and lessen risks and aggravations to the wellbeing in situation of crisis, in this way enhancing the personal satisfaction in the Commonwealth (Department of Health Commonwealth of Virginia, 1984). In the United States, responsibility for public health rests with the states and local authorities. This is mainly because every state is different in its geography, politics, nature of population, health system and challenges. Four functions of the states' public health system are: regulation, financing, resource allocation and provision of services For regulating these core functions, a state has many more participants than the global health system such as, state legislatures, county commissions, the business community, employers, media, and advocacy and constituent groups (Beyle and Dusenbury, 1982) (Table 1.2).

TABLE 1.2   Assessment Activities of State Health Agencies, 1984

Data Collection

	Number of States (n = 46)
Vital records and statistics	44
Morbidity	24
Health facilities	39
Health manpower	38
Hospital care	32
Ambulatory care	19
Long-term care	28
Health systems funds	22
Health interview surveys	20

Source: From Institute of Medicine, Division of Health Care Services, Committee for the Study of the Future of Public Health. 1988. The Future of Public Health. Washington, D.C.: National Academy Press. Retrieved from https://books.google.co.id/books?id=pHDt0C3LEagC&pg=PA173&lpg=PA173&dq=American+Medical+Association,+Department+of+State+Legislation,+1984&source=bl&ots=Jh2k24bxcP&sig=ACfU3U3lfQePyBBNLxND_0_UpRBIpY450g&hl=en&sa=X&ved=2ahUKEwjLvqHej5jhAhXEfn0KHaD7DSoQ6AEw.

Depending on the organization of the state health authority, it will relate differently to these various federal agencies.

Local/Decentralized: Local health departments play major roles in these situations and make major decisions.

Mixed: In this situation, some health departments are led by local government and others by state government.

State/Centralized: In this case, all local health departments are ruled by state units.

Shared: In this case, all local health departments' responsibilities are shared by local and state units. (Council of State Governments, 1987).

Public health education is mainly started in 1918 with the establishment of the Johns Hopkins University of Hygiene and Public Health. According to the Council on Education for Public Health, in the next 10 years the number of public health schools may double. Apart from epidemiology and biostatistics, new fields are emerging such as public health informatics, management of clinical trials and human genetics. With public health ascending high on the national plan and a wealth of assets being guaranteed, maybe there is currently a chance, as there has not been for an exceptionally prolonged stretch of time, to shape a future arrangement of general well-being instruction that tends to the issues that have been so regularly depicted and examined. If we talk about the role of a public health educator, it starts with exploring the needs of a community by enquiring about their suffering because of lack primary health care knowledge and providing them with resources. After the phase of exploration then come promotion, implementation, evaluation, management and advocacy of health, and health education. A health advocator is responsible to foster the healthcare profession in the community and works with other people to regulate the standards and to achieve the highest level of health outcomes.

Public health systems, as noted earlier, exist on the national, state, and local levels. There are numerous channels for data and facilitated action among national, state, and local levels in both public and private sectors, similarly as there is trade of data and coordination of action between the health care system, different offices, and private entities. The framework involves both different levels of government and different organizations.

CHAPTER 2

Continuing Education in Public Health in the United States

2.1 INTRODUCTION

The American Public Health Association (APHA) is a primary public health advocate protecting all Americans and their communities from preventable, serious health threats including emergency preparedness, food safety, hunger and nutrition, climate change and other environmental health issues, public health infrastructure, disease control, international health and tobacco control. The APHA represents a broad array of health providers, educators, environmentalists, policymakers and health officials that strive to ensure that community-based health promotion, disease prevention activities, preventive health services are accessible to everyone, providing funding for core public health programs, and eliminating health disparities.

2.1.1 Historical Perspective

The APHA brings together various health professionals such as physicians, nurses, pharmacists, epidemiologists and public health educators. The APHA, being a tremendous asset to public health, is known as the only organization with the ability to influence federal policy for securing funds for health programs, awareness, or emergency preparedness to improve public health.

The APHA plays a critical role in strengthening health professions across health fields by providing continuous public health education, enhancing skills in public health practice, theory, research and policy. Public health in many instances is difficult to define and often people are clueless about public health. A telephone survey among 1234 registered voters conducted in 1999 by a charitable foundation found that over half of the respondents misunderstood the term.

In 1920, Charles-Edward A. Winslow defined public health as, "the science and the art of preventing disease, prolonging life, and promoting physical health and efficiency through organized community efforts for the sanitation of the environment, the control of community infections, the education of the individual in principles of personal hygiene, the organization of medical and nursing services for the early diagnosis and preventive treatment of disease, and the development of the social machinery, which will ensure to every individual in the community a standard of living adequate for the maintenance of health"(Winslow, 1923).

The many successes of public health have dealt with solving precise issues such as championing laws that promote smoke-free indoor air and wearing seatbelts, promoting a healthy lifestyle, and bestowing evidence-based solutions to problems described in Winslow's definition. The APHA's overall mission is strengthening preventive services and prompting wellness by encouraging healthy behaviors. According to the APHA, public health endeavors include disease surveillance, research on disease demographics and prevention of catastrophes.

The APHA accredits public health professionals as pioneers for public health advocacy. The APHA's policies are based on scientific research and member-led processes. The APHA "Speak for Health" initiative provides one with the tools to deliberate important public health issues either by meeting with or calling members of Congress or their staff, attending a public forum, and presenting opportunities for speaking at a town hall or public meetings. Furthermore, it provides action alert messaging and phone scripts via text message or by phone.

2.2 ROLE OF APHA IN CONTINUING PUBLIC HEALTH PROFESSIONAL DEVELOPMENT

Continuing education (CE) assists professionals in maintaining their professional license, and CE credits are often required by an employer or even by the government. Course curriculum evolves as new advances are being made in health care and CE helps to ensure that professionals stay informed. CE has been shown to improve patient outcomes by continuously advancing on patient care. The sole purpose of CE became critical in health care back in the 1970s, due to the advances in new technologies and developments. Now, new specializations are emerging such as human genetics, clinical trials, and public health informatics which help in advancing education.

Health care professionals are more mindful of the issues than their 1970s counterparts because of the advances in research and technology. Many schools and competing organizations across the globe are engaging in distant learning programs through the Internet that offer the possibility of fulfilling the long-recognized need to bring public health education to the homes and offices of the public health workforce.

2.3 CONTINUING EDUCATIONAL PROGRAMS UNDER APHA

The APHA offers the APHA Integrated Continuing Education Program for public health professionals and those interested in public health. The main goal of the online program is to provide career and professional development opportunities through education; easy access to expanded opportunities; webinars on relevant topics; and links to relevant resources, meetings and activities that might be of interest.

2.4 CONTINUING EDUCATION IN PUBLIC HEALTH MISSION AND ACCREDITATION IN THE UNITED STATES

The APHA has a structured set of policies, with a mission statement that aims to expand one's current skills or knowledge. In order to improve the health of the public and health care system, a set of guidelines must be followed. The APHA provides quality continuing education (CE) for public health and health care professionals. The APHA's mission approves education activities that are developed externally. It is important to note the Association's CE program is multi-disciplinary with a single set of policies and processes that meet the requirements of the accreditations that it holds.

Public health–related CE is a learning experience designed to augment the knowledge, skills, competence, performance, and attitudes or the professional development of the workforce. Such learning is aimed at improving the public health and the health care delivery system by presenting evidence-based practice and practice-based evidence in such contexts as public health education, policy, regulation, law or other relevant environment. A number of public health professionals in other disciplines may also benefit from these accreditations for re-licensure, re-certification or other recognitions. If CE credit hours have been awarded, it is up to the professional to determine if they can apply them to their re-licensure, or re-certification.

The Association's priority is to ensure compliance with the following CE accrediting organizations:

1. National Commission for Health Education Credentialing, Inc., for certified health education specialists;

2. Accreditation Council for Continuing Medical Education, for physicians and non-physicians;

3. American Nurses Credentialing Center Commission on Accreditation, for nurses; and

4. National Board of Public Health Examiners, for certified public health professionals

The responsibility for compliance with accreditation requirements for CE is that the activities are shared among all APHA planners, faculty/presenters, panelists, moderators, authors, planning reviewers (referred to as content reviewers or content experts), and the CE staff in the APHA Learning Professional Development Programs Unit with a role of developing educational opportunities that are based on adult learning principles. CE is used to build on one's basic professional education, to stay updated, and to acquaint the learners with the practical knowledge that can be clinically applied. Adult learning principles are also applied when developing the content and selecting the appropriate teaching method/strategy. (Knowles, 1973, 1990; Knowles, 1984; Jarvis, 1985; Merriam and Caffarella, 1991; Senge, 2006; Merriam, 2001).

The expected results from the multi-disciplinary CE program include: (1) increase the ability of public health practitioners to identify, adapt and apply strategies to implement and/or integrate new technologies such as health information technology (IT) into their practice; (2) escalate public health practitioners' ability to identify, adapt and apply strategies to integrate and implement evidence-based approaches in addressing social determinants of health and other factors that affect patient outcomes; and (3) increase public health practitioners' ability to identify, adapt and implement population based health strategies to increase the health status of their target audience.

The APHA holds an annual meeting to enhance the knowledge of health professionals and practitioners. The APHA provides information exchange on best practices, latest research, and new trends in public health.

Objectives of the APHA are as follows:

- Addressing the gap between public health practitioners' knowledge and performance as it relates to developing and implementing policies that demonstrate a collaborative approach to improving the health of all people and advancing health equity.

- Increasing the knowledge of the public health community to incorporate health equity, and sustainability principles into specific policies, programs, and processes.

- Identifying best practices in developing and implementing strategies that address the social determinants of health.

- Improving the competencies of public health practitioners through skill building.

2.5 PROFESSIONAL DEVELOPMENT

Some of the professional development opportunities that are offered to public health professionals include: Public Health Careermart Continuing education (CE), fellowships, mentoring, and internships which are all aimed at preparing and helping the public health professionals excel.

Public health Careermart is a platform where the employees and employers come together and provides the following services:

- Career Learning Center
- Career Coaching
- Resume Writing
- Reference Checking

The CE program aims at career, education, and professional development for public health practitioners. These credits are offered online or through the conferences and annual meetings for APHA.

The APHA CE component of its annual meeting is compliant with the accreditation criteria of each of the following accrediting bodies. For further information, refer to the printed annual meeting program tab labeled "Continuing Education," or contact CE Services at 202-777-2491.

Continuing Education	Type Accrediting Body
Certified Health Educators[a]	National Commission for Health Education Credentialing, Inc.
Physicians[b]	Accreditation Council for Continuing Medical Education
Other healthcare professionals[c]	All other health professionals will be awarded certificates of attendance for CME designated for AMA PRA Category 1 credit TM.
Nurses[d]	American Nursing Credentialing Center Commission on Accreditation
Certified Public Health Professionals[e]	National Board of Public Health Examiners

[a] Certified Health Education Specialist (CHES® and MCHES®): The APHA is accredited by the National Commission for Health Education Credentialing, Inc., to sponsor continuing education for Certified Health Education Specialists and Master Certified Health Education Specialists.

[b] Continuing Medical Education (CME): The APHA is accredited by the Accreditation Council for Continuing Medical Education to provide continuing medical education for physicians.

[c] Designation Statement: It is the statement to confirm individual confirmation for any act. APHA designated this educational activity for a maximum of 31.5 AMA PRA Category 1 Credit(s)TM. Physicians should only claim credit commensurate with the extent of their participation in the activity.

[d] Continuing Nursing Education (CNE): The APHA/PHN Section is accredited as a provider of continuing nursing education by the American Nurses Credentialing Center's Commission on Accreditation.

[e] Certified in Public Health (CPH): The APHA is an approved provider of Certified in Public Health Continuing Education Credit (CPHCE) by the National Board of Public Health Examiners.

For CE accreditation compliance, view the APHA Continuing Education Policies. Continuing education information is online and in the hardcopy program. See the "Continuing Education" tab in the final program or the online program for more information.

Please note that the final online program includes:

1. Meeting purpose, learning outcomes and practice gaps

2. Agenda

3. Conflict of interest resolution policy

4. Commercial support statements

5. Abstract/proposal

6. Presenter, planner, moderator and organizer disclosures

7. Non-endorsement statement

Please take advantage of new Web technology to plan ahead. Try the "Personal Scheduler" to help you map out the sessions and events that you would like to attend, print out individual abstracts, and use the powerful search engine to help you identify sessions of interest. The online program can also be accessed on site from the Electronic Information Centers at the Convention Center.

2.5.1 Online Continuing Education Program

The APHA Center for Professional Development offers CE credits through its online program. Individuals can earn CE credits from the comfort of their home or office and at their convenience. It simply requires listening to the recorded sessions, passing the online quiz, completing the session evaluation and printing the certificate.

Upcoming opportunities for CE credits include the APHA Annual Meeting scientific session recordings, APHA webinars, and APHA-developed workshops. Sessions are added continuously. As part of APHA's commitment to provide access for persons with disabilities, closed captioning and/or transcripts are provided for all webinars.

Types of credits offered are as follows:

- CHES®: Certified Health Education Specialist

- CME: Continuing Medical Education

- CNE: Continuing Nursing Education

- CPH: Certified in Public Health

- OP: Other Professional (check with your licensing or certification board to see if they accept CME for non-physician credits)

2.6 POLICIES FOR THE CONTINUING EDUCATION FOR LEAD CE PLANNERS

The eligibility of becoming a Program Planner is decided by the lead CE planners. Their responsibilities include reviewing the completed Biodata/Conflict of Interest (COI) form from each planner and making decisions based on being free of bias or having no conflict of interest, whether potential or actual, and signing off for one to become a Program Planner. If a potential bias or conflict of interest is identified, it must be resolved before that individual can proceed to the Program Planner role. Documentation of resolution and method used with the CE planner's signature is also required.

2.6.1 Eligibility to Become a Planning Reviewer

Before the Planning Reviewers (or content reviewers or content experts) begin actual responsibilities, the lead CE planners must review the completed Biodata/COI form from each planner and decide if that individual is free of bias or conflict of interest, whether potential or actual, and must sign off for one to become a Planning Reviewer.

2.6.2 Policies for Program Planners and Faculty/Presenters of the APHA Annual Meeting or Other Meetings

The Program Planners are those individuals who are identified as the planners for each section, special primary interest group (SPIG), forum, caucus, student assembly, and the APHA boards and committees which develop scientific sessions. It is mandatory for Program Planners to attend the APHA Planning Orientation annually and usually work with a group of individuals who help with the planning. The other individuals are referred to as Planning Reviewers (or content reviewers or content experts). The APHA planners and the potential faculty/presenters develop the educational content of the scientific sessions and the Learning Institutes.

Program Planners work with the Planning Reviewers to ensure that:

1. Potential faculty/presenters qualify to present the content that they are going to propose.

2. Content proposed is relevant for the meeting theme and meets the learning needs of the target audience who are attending the meeting.

3. Individuals who submit abstracts follow all the instructions that are in the Conference and Abstract management software (CONFEX) system.

At the annual meeting, the Program Planners and Planning Reviewers use an objective rating form at the time of evaluating the submitted abstracts for acceptance or rejection for presentation. The APHA has an evaluation form that is present on the CONFEX online system. In addition to the standard form, some components that develop scientific sessions may also use a customized review form that includes the interest of a specific component.

In general, the responsibilities for developing the following elements of the educational component for the APHA annual meeting are the following:

- Overall meeting theme—APHA Governing Council.

- Purpose—Lead CE Planners for each of the accredited disciplines.

- Learning needs assessment—APHA Education Board CE Committee.

- Identification of at least one gap in knowledge, skill or competence, performance or professional development that is based on the findings of the needs assessment—Lead CE Planners for each of accredited disciplines.

- Development of the Calls for Abstracts for the Overall Meeting—APHA Program Planners-at-Large.

- Development of the Calls for Abstracts are specific to each component in APHA that plans scientific sessions—Program Planner and Planning Reviewers for that component (i.e., section, SPIG, caucus, forum).

2.6.3 Responsibilities of the Faculty/Presenters

Each faculty/presenter who submits an abstract has the following responsibilities. The faculty/presenters may interact directly with their respective Program Planner or Planning Reviewer through either CONFEX system, email or by phone.

Policy 1. Educational Content and Content Integrity
The 2010 Institute of Medicine (IOM) report noted that educational activities must be conducted with integrity, ensuring freedom from commercial bias or promotion and based on best available evidence. The educational session/activity must be independent, with well-defined objectives, and free from bias by a professional, financial, personal, or a commercial interest. The Program Planners may need to work with potential faculty/presenters to assist them in bringing a presentation into compliance with the requirement for content integrity.

2.6.4 Content and Educational Design Components

- Identification of CE learning needs: The learning need underlies a professional practice gap in knowledge, competence in practice or evidence-based practice. The APHA conducts a learning need assessment for public health professionals. These findings are incorporated into the overall Annual Meeting Call for Abstracts that the Association's program planners publish. The planners from the various APHA components do not need to conduct an additional needs assessment. However, some components choose to identify the learning needs for their specialty component. If this happens, the findings are expected to be incorporated into Abstracts.

- Identification of a professional practice gap that the presentation will address. Choices are listed in the CONFEX online submission screens.

- Determination of the title of presentation that reflects the topic that is to be presented.

- Determination of at least one measurable learning objective per abstract submission. Each objective must have only one action verb. The APHA provides a list of acceptable action verbs in the CONFEX online system. The APHA only evaluates the first listed learning objective. The Learning Institutes may have additional learning objectives depending on the length of the activity, the number of speakers and the content. A list of acceptable and of non-acceptable objectives is in the CONFEX online submission system on the screen to enter the learning objective for each presentation.

- Selection of content/abstract or outline. An Abstract should be at least two sentences that explain/describe the presentation that is proposed. It must have a well-defined objective, free from bias and promotion. It may not include the names of commercial entities, products or services. Using classes of drugs or generic names of drugs and medical devices is acceptable.

- Selection of faculty/presenters, panelists, responders and moderators. Individuals in these roles either develop the content, learning objectives, and teaching strategies, or they participate in the development, or respond in their own way to what is developed by the lead faculty/presenter or keynote speaker. They all have the ability to influence the content as it is presented to the learners, although how they may influence will vary depending on each of these roles and the choice of the individual. Each presenter/faculty must complete a Qualification Statement that reflects one's ability to deliver appropriate content. The Program Planners ensure that the faculty/ presenters are qualified by education and/

or experience to present the specific content that is in the abstract.

- Selection of educational strategy/method, materials and resources such as handouts reflect the application of adult learning principles. Evidence supports that active engagement in the learning process is key for adult learners. (Benner et al., 2010; Senge, 2006). The learning method is based on how best to convey the content given the needs assessment and the gap that is being addressed. Acceptable strategies include presentation followed by question and answer; keynote followed by discussion; panels with discussions; and/or demonstrations, interactive learning, small group activity and/or roundtable discussion.

2.6.5 CE-Eligible Content

CE-eligible content aims at improving practice by enhancing one's knowledge, skill, competence, performance or professional development. The ultimate intent is to improve health and health care for all people. In public health, scope of the content topics that are acceptable for CE include health care systems and programs; delivery of health care; health-related policy, regulation, laws, funding, and other topics that keep the professionals updated in their specialty. CE-eligible content should be based on the best available evidence or experience. Evidence-based practice and practice-based evidence for health professionals are the most preferred types of content. The content of a presentation is congruent with the presentation's title, identified gap and learning objective. The content is appropriate for the suggested target audience. The APHA need assessment applies to all of the sessions and presentations that are selected for the annual meeting.

2.6.6 CE-Ineligible Content

Types of content that are not eligible for awarding CE contact hour credits include: management of personal finances and investments for target audience; content that promotes a product, service or other entity in a preferential or biasing manner; advocating, lobbying; political campaigning; rallies; strikes; and other actions that aim to influence or promote actions in a particular direction.

2.6.7 Content Integrity

Content integrity is the development of a learning presentation that is based on objective evidence and/or experience. Objectively presenting evidence that is positive or negative on a topic is acceptable. The basis for positives or negatives needs to be given in a factual manner and supported with evidence where possible. Comparisons and contrasts need to be factual in order to appropriately inform the audience and protect content integrity. Content integrity is free of bias.

2.6.8 Bias

Bias may be described as showing prejudice or personal preference that may influence what content is presented or the manner of presentation. It is truly interjecting preferences or attempting to influence the learners toward one bias, presenting the pros and cons or the positive effects as contrasted with potential downsides of a product, service, program, approach to care, policy or other topic.

2.6.9 Measuring Outcomes

The lead CE planners for the annual meeting review and modify the Learner Evaluation Form. The form is available for all Program Planners to view. If a particular subgroup of the target audience wishes to conduct an additional evaluation from the learners, that subgroup is allowed to do so. However, the learner evaluation that the CE planners develop is the one that must be completed and submitted as one of the requirements for an individual to earn CE contact hour credits.

One of the approaches to outcome measurement is to evaluate whether the activity helped in narrowing the gap between the learners' current knowledge or abilities and what was desired.

Selection of a desired outcome is purposeful in nature and drives development of objectives, content, and teaching/learning strategies. The learners will be informed of the intended outcome and the criteria for successful completion prior to the beginning of the learning activity. The evaluation feedback is analyzed and incorporated as one of the sources of input for the learning needs assessment of future programs.

Policy 2. Policy on Conflicts of Interest

Program Planners and Planning Reviewers should assess each accepted submission for a conflict of interest (COI). All Program Planners, Planning Reviewers, faculty/presenters, panelists, respondents, and moderators are required to complete a Biodata/COI Form that is on the CONFEX online system where the names of individuals in these roles are entered. The faculty/presenter, panelist, responder, or moderator could have a COI. Also these individuals need to declare if a spouse or partner (personal or professional) may have a relationship that is defined as a COI.

A COI is an affiliation or relationship of a financial nature with a commercial interest organization or entity whose products or services are consumed by patients, and these are included in the educational content of the presentation. Such a relationship constitutes a COI because it might bias a person's ability to objectively participate in the planning, implementation, or review of a learning activity. A COI may be actual or potential. If a reasonable person might perceive a COI, then it is a perceived conflict. An actual COI exists when one has a financial, professional and/or personal relationship with a commercial interest entity.

The types of relationships that may constitute a COI are categorized in the list below. An individual may have more than one type at the same time

- A "financial interest" may include, but is not limited to, a financial benefit that is expected by an individual through employment such as a wage or salary, self-employment, independent contractor, an intellectual property right that results in a royalty or other remuneration, consulting or speaking fee, teaching pay, honoraria, ownership interest (e.g., stocks, stock options, or other ownership interest, excluding diversified mutual funds), membership on advisory committee, review panel, board, or other activity from which remuneration is received or expected.

- A "professional interest" may include, but is not limited to, a situation in which an entity receives a contract or grant and manages the funds, but an individual is the principal, named investigator, or is in any position to influence the results or outcomes. This includes students.

- A "personal interest" may include, but is not limited to, a financial relationship that is held by one's spouse or partner. Also any of the relationships mentioned above may also be a "personal interest."

A COI must be disclosed for an ongoing conflict and for a period of 12 months after it has been resolved. Defiance to disclose the conflict disqualifies one from participating in CE granting activities.

A commercial interest organization or entity is described as any entity either producing, marketing, reselling, or distributing health care goods or services consumed by or used on patients or an entity that is owned or controlled by an entity that produces, markets, resells, or distributes health care goods or services consumed by or used on patients and the content of the educational activity/session/institute or presentation includes these products or services.

Entities that are exempt from being a commercial interest entity are nonprofit or government entities and other non-health-care related companies.

The APHA is required to have a process to identify and resolve any COI before the CE activity is presented. This form requires disclosure of any financial, professional or personal conflict and it requires a signature to agree to follow the APHA policies that resolve the conflict. The process includes:

- Planners, presenters, authors, moderators, panelists, or speaker respondents must complete an APHA COI form before assuming these roles.

- A completed COI form is available via website when it is opened up to all attendees to view which is weeks before the activity. Disclosure of whether a COI may or may not exist is included in the APHA annual meeting program, both online and in a hard copy.

- Exception: When there is a last minute change in presenter, moderator, panelist, or respondent, then this is announced at the beginning of the educational session and a completed COI form is submitted to the CE staff as soon as possible either when the change happens, or immediately after the session.

- The APHA meets the requirements for disclosure to activity/ session learners through the advance publication in the online APHA program. No verbal disclosures are required at the time of the presentation. However, presenters or moderators may opt to give an oral disclosure but this is not a requirement and documentation is not required.

Policy 3. Commercial Support
Commercial support is defined as financial contributions given by a commercial interest entity, in the interest of paying for some or all of the costs of the educational activities.

The planner should contact the CE staff to follow the appropriate process of accepting commercial support related to learning institute, roundtable scientific and theater sessions.

Use of section enrichment funds to provide remuneration to speakers must be reviewed by the staff of CE to ensure compliance.

The APHA accepts the commercial support for learning institutes, scientific session, roundtable or theater, in accordance with ACCME Standards of Commercial Support and ANCC Content.

Policy 4. Sponsorship
Sponsorship is a financial contribution from an organization when the organization does not fit into the category of a commercial interest.

When a provider of a CE activity joins with another entity and that entity contributes to the finances, then the relationship becomes a sponsorship.

In that case, the APHA remains the provider and the other organization is called the sponsor.

For the APHA annual meeting, any contribution must be processed through the Association and the APHA sponsorship form must be completed. Any funds contributed must be submitted to the APHA and the Association must distribute the funds according to the sponsorship form designations.

Contributions that are given to the APHA or to one of the sections must be documented; they might be marketing and event registration assistance or a meeting room. All such contributions to the Association and its components must be processed by the APHA and a sponsorship form must be completed.

2.7 HISTORY AND CURRENT STATUS OF PUBLIC HEALTH EDUCATION IN THE UNITED STATES

2.7.1 History
This section discusses two broad phases of public health education in the United States:

1. The first phase (roughly between 1914 and 1939), during which self-reliant public health schools were first established which were privately funded by philanthropies.

2. The second phase (from 1935 to the present), slightly overlapping with the first, was marked by federal and state funding.

2.7.2 Public Health Education: 1914–1939

By the end of the 19th century, medical and nursing schools had flourished and were mainly established by hospitals to provide a source of well-trained labor. However, there was no distinct education or career path for public health officers; most of them were practicing physicians who were called upon in times of crisis. It was in this context that the staff of the Rockefeller Sanitary Commission (RSC) attempted to enlist public health officers in the southern United States to aid in a campaign to eradicate hookworm. They found little interest in public health, leading Wickliffe Rose, the architect and organizer of RSC, to believe that a new profession was needed, composed of men and women who would devote their entire careers to controlling disease and promoting health at a population level. Three possible approaches for public health education were debated—the engineering or environmental, the sociopolitical, and the biomedical.

Rose with Abraham Flexner established a separate public health career. On October 16, 1914, Flexner brought together 11 public health representatives and 9 Rockefeller trustees and officers for a meeting. It was decided that there were essentially three categories of public health officers: those with executive authority, such as city and state health commissioners; the technical experts in specific fields such as bacteriologists, statisticians, and engineers; and the field workers such as local health officials, factory and food inspectors, and public health nurses.

Rose laid out ideas for a system of public health education centered on a university affiliated with research incentive, separate from a medical school, whose graduates would be strategically placed throughout the United States. This central scientific school of public health would be linked to a network of state schools that sent extension agents into the field, and emphasized not only public health education, but offered short courses and extension courses to upgrade the skills of health officers in the field, and also demonstrations best practices. This plan only focused on research, ignoring public health practice, administration, public health nursing, and health education. The biomedical side of public health was emphasized to the exclusion of its social and economic context and no attention was paid to the political sciences or to the need for a social or economic reform.

The Johns Hopkins School of Hygiene and Public Health became the first endowed school of public health, during the influenza epidemic of 1918. Later, the Rockefeller Foundation officials agreed to provide funding for additional schools of public health including ones at Harvard and Toronto. These first schools were well-endowed private institutions that favored persons with medical degrees, had curricula that leaned heavily toward the laboratory sciences, and emphasized infectious diseases and granted fellowships to medical graduates worldwide, but lacked emphasis on field training. By 1930 these first schools were graduating a small cohort of individuals with sophisticated scientific education but they were not producing the needed large numbers of public health officers, nurses, and sanitarians.

2.7.3 Public Health Education: 1935 to the Present

Passage of the Social Security Act of 1935 provided a major stimulus for further development of public health education. Provisions of this Act increased funding for the public health services and provided federal grants to the states to assist them in developing their public health services. Federal law now required each state to establish minimal qualifications for health personnel employed using federal assistance, and recommended at least one year of graduate education at an approved school of public health. Overall, the states budgeted for more than 1,500 public health trainees, and the existing training programs were soon filled to capacity. As a result of the growing demand for public health credentials, several state universities began new schools or divisions of public health and existing schools of public health expanded their enrollments.

In 1936, 10 schools that offered public health degrees or certificates requiring at least one year of residence were Johns Hopkins, Harvard, Columbia, Michigan, University of California at Berkeley, Massachusetts Institute of Technology, Minnesota, Pennsylvania, Wayne State, and Yale (Committee on Professional Education, 1948). By 1938, more than 4,000 people (1,000 of whom were physicians) had received some public health training. Increased funding and the continuing need for additional public health graduates led many colleges and universities to inaugurate public health departments. Federal training funds were allotted to universities in California, Michigan, Minnesota, Tennessee, and North Carolina to develop short courses for the rapid training of public health personnel.

The tremendous push in the late 1930s toward training larger numbers of public health practitioners accelerated practical training programs rather than research. Public health departments wanted personnel with one year of public health education; typically, the Master of Public Health (M.P.H.) degree or an individual with a few months of public health training. Ideally, persons who understood practical public health issues rather than scientific specialists with research degrees were accepted. Thus, public health education in the 1930s was more practically oriented, with considerable emphasis on fields such as public health administration, health education, public health nursing, vital statistics, venereal disease control, and community health services. During this period, many schools developed field training programs in local communities where their students could obtain experience in the practical world of public health and prepare for roles within local health departments. Thus, the 1930s was the prime era of community-based public health education.

The growth of short training programs in public health education continued throughout the initial years to meet the demand for physicians, nurses, and sanitarians with at least minimal training in tropical diseases, parasitology, venereal disease control, environmental sanitation, and a variety of infectious diseases. For the burgeoning industrial production areas at home, industrial hygiene was in demand; for areas with military encampments, sanitary engineering and malaria control were urgent concerns.

Schools of public health and public health training programs revamped their educational programs to meet the needs and turned large numbers of health professionals with a smattering of specialized education in high-priority fields. The research-oriented schools of public health, such as Johns Hopkins and Harvard, maintained their research programs largely by recruiting foreign students—many from Latin America—to staff their laboratory and field programs.

In 1941, the Association of Schools of Public Health (ASPH) was founded to promote and improve graduate education for public health professionals. In 1946, the Committee on Professional Education of the American Public Health Association began monitoring the standards of public health education amid complaints that profit-making public health training courses of questionable quality were being offered by correspondence from faculty who did not even know of their appointment. A 1950 survey of schools of public health found major difficulties such as these schools being overcrowded, underfunded, and lacked key faculty members, classroom and laboratory space, and necessary equipment (Rosenfeld et al., 1953).

Given the high demand for public health graduates and the need for schools and programs to train them, it is not surprising that the criteria for accreditation of schools of public health as implemented at mid-century were relatively undemanding. To become accredited, schools were required to have at least eight full-time professors as well as lecture rooms, seminar rooms, and adequate laboratory facilities; and they were to be located close to local public health services that were of sufficiently high quality that could be used for "observation and criticism to make observation fruitful" (Winslow, 1953).

For a few years following World War II, the concepts of social medicine, social epidemiology, and the ecology of health achieved prominence. Schools of public health developed new courses that focused on world population and the food supply; the impact of industry and transportation on health; the impact of cultural, social, and economic forces on health; evaluation of health status; and public health as a community service (Winslow, 1953). At the University of Pittsburgh, Thomas Parran, a public health service officer and physician, had decided that the curriculum should be organized around "the systematic presentation of illustrative topics which deal with the interrelation of man and his total environment and with the political, economic, and social framework within which the health officer must work" (Blockstein, 1977). Yale University's core course on "Principles and Practice of Public Health" was similarly organized around a series of interdisciplinary seminars running throughout the academic year. Winslow commented approvingly that the eleven schools of public health constituted "eleven experimental laboratories in which new pedagogic approaches are constantly being devised" (Winslow, 1953).

The overall impression of the accredited schools of public health in 1950 was that they were doing a good job of preparing public health practitioners through courses and fieldwork, that the numbers of faculty and students were growing, and that curricular and research innovations seemed promising. The main complaints of the schools seemed only to be lack of funding.

Schools of public health were concerned about lack of money as they were competing with major funding with medical schools. The finances were allocated towards construction of community hospitals through the Hill-Burton Program, and the National Institutes of Health (NIH) was experiencing rapid growth in research funding. The NIH expanded with enormous increases in financial resources, transferring most of its funds to universities and medical schools in the form of research grants. Grants were awarded based on the decisions of peer review committees composed of non-federal experts in the relevant fields of research.

Liberals, conservatives, medical school deans, and researchers were all happy with the system, and members of Congress were pleased to bankroll such a popular and uncontroversial program (Strickland, 1972; Ginzberg and Dutka, 1989).

To survive, schools of public health turned to research funding to pay the salaries of additional faculty members, using the rationale that new faculty could focus on teaching as well as on research. As this strategy was implemented, a particular department within a school was devoted mainly to teaching or to public health practice, the number of faculty employed increased or decreased. If the department was devoted to research and was reasonably successful at funding that research, the department flourished. Even the schools that strongly favored teaching and field training over research became unable to resist the pressures that encouraged research over practical training. Available funding, and faculty who were suited by education, experience, and personality to succeed in the research system, shaped the schools of public health. Because the system of research funding was not oriented toward field research, public health practice, public health administration, the social sciences, history, politics, law, anthropology, or economics, the laboratory sciences tended to thrive while the practice and other non-quantitative disciplines suffered. The community-based orientation of the 1930s disappeared, and the field training programs virtually ceased to exist.

As faculty engaged in their laboratories, they further distanced themselves from the problems of the local health departments, which were experiencing increasing difficulty. Federal grants-in-aid to the states for public health programs steadily declined during the 1950s as the total dollar amounts fell from $45 million in 1950 to $33 million in 1959. Given inflation, this represented a dramatic decline in purchasing power (Terris, 1959). Lacking funds, health departments could not afford new people or initiating new programs. Health departments ran underfunded programs with under qualified people who answered to unresponsive bureaucrats.

Between 1947 and 1957, the number of students educated in schools of public health fell by half. Alarmed, Ernest Stebbins of Johns Hopkins and Hugh Leavell of Harvard, representing ASPH, urged Congress to support public health education. They found an especially sympathetic audience in Senator Lister Hill and Representative George M. Rhodes, and in 1958, Congress enacted a two-year emergency program authorizing $1 million a year in federal grants to be divided among the accredited schools of public health.

The First National Conference on Public Health Training in 1958 noted that these funds had provided 1,000 traineeships and had greatly improved morale in public health agencies. The conference further requested appropriations for teaching grants and construction costs for teaching facilities, and urged that faculty salary support be provided for teaching. Its report concluded with a stirring appeal to value public health education as vital to national defense. This depends on specially trained physicians, nurses, biochemists, public health engineers, and other specialists properly organized for the normal protection of the homes, the schools, and the workplaces of some unidentified city somewhere in America.

President Dwight Eisenhower signed the Hill-Rhodes Bill, authorizing $1 million annually in formula grants for accredited schools of public health and $2 million annually for five years for project training grants; between 1957 and 1963 the United States Congress appropriated $15 million to support public health trainees. Between 1960 and 1965 the total number of applicants to schools of public health more than doubled; the number of faculty members increased by 50%; the average space occupied increased by 50%; and the average income of the schools more than doubled (Fee and Rosenkrantz, 1991). Following the passage of Medicare and Medicaid legislation in 1965, state health agencies turned to schools of public health to provide the scientific basis for rational decision-making in health services delivery and training for medical care administrators and financial managers. ASPH estimated that 6,220 new positions in medical care administration required graduate-level education (ASPH, 1966).

The 1960s brought major progress for the civil rights movement and for President Lyndon B. Johnson's War on Poverty which included the Office of Equal Opportunity (OEO). The OEO helped create 100 neighborhood health centers and the Department of Health, Education, and Welfare (DHEW) supported another 50. A strong environmental movement developed following the publication of Rachel Carson's *Silent Spring* in 1962. In 1970 Earth Day attracted 20 million Americans in demonstrations against assaults on nature; by 1990 Earth Day brought out 200 million participants in 140 countries (McNeil, 2000). The Environmental Protection Agency (EPA) was established and the first Clean Air Act was passed in 1970. Also created during this period were the Occupational Safety and Health Administration (OSHA) and the National Institute of Occupational Safety and Health (NIOSH).

Throughout the 1960s and early 1970s, schools of public health thrived with federal funding available for both teaching programs and research. In 1960 there were 12 accredited schools of public health in the United States, 8 were added between 1965 and 1975. Between 1965 and 1972, student enrollments again doubled, with the large majority being candidates for the Master of Public Health (M.P.H.) degree. The trend continued to admit more students who were not physicians, and more students without prior experience in public health. In 1946, 61% of all students admitted to schools of public health for the M.P.H. were physicians; by 1968–1969 that figure had dropped to only 19% of M.P.H. candidates.

Along with the growth in the accredited schools of public health came a rapid growth in other forms of public health and health services education. Graduate programs were established in a variety of university departments and schools (e.g., engineering, medicine, nursing, business, social work, education, and communication) offering degrees in such fields as environmental health, health management and administration, nutrition, public health nursing, and health education. Universities were creating popular baccalaureate programs in health administration, environmental engineering, health education, and nutrition. By mid-1970, some

69,000 students were enrolled in various allied health programs (Sheps, 1976). Although 5,000 graduate degrees in public health were awarded each year, approximately half of higher education for public health was occurring outside of accredited schools of public health.

Then, in 1973, President Richard M. Nixon recommended terminating federal support for schools of public health and discontinuing all research training grants, direct traineeships, and fellowships. J. Thomas Grayston of the University of Washington reflected the thoughts of the field when he said: "the greatest immediate challenge to the School of Public Health and Community Medicine is the uncertainty of federal funding brought about by the administration's announced intention to end federal support for the training of public health manpower, coupled with a similar proposal to decrease support for research training" (Grayston, 1974).

The threatened elimination of funding was averted, however, and in 1976 Congress passed the Health Professions Educational Assistance Act (P.L. 94–484), which provided for a number of programs in health professions education. The trend, however, was toward ever more reliance on targeted research funding. Also in 1976 the Milbank Memorial Fund issued its extensive report, Higher Education for Public Health, proposing a new structure for the public health educational system—a three-tiered structure.

First, schools of public health were to educate people to assume leadership positions. Next, programs in graduate schools would prepare the large number of professionals engaged in providing clearly differentiated specialty services, for example, public health nurses, health educators, and environmental health specialists. Finally, baccalaureate programs could provide some of the "trained entry-level personnel" (MMF, 1976). The report identified three core areas of public health on which the schools should focus: epidemiology and biostatistics, social policy and the history and philosophy of public health, and management and organization for public health. In addition, the report recommended that schools should serve as regional resources by helping faculties in medical and other health-related schools to develop teaching programs and research in public health; they should become involved in the operation of community health services; and schools should design their research within a broad framework established by the needs of public health practice.

Under President Ronald Reagan the pressures intensified. Between 1980 and 1987, spending for health professions' education by the Department of Health and Human Services (DHHS) Bureau of Health Professions declined annually by more than 50% from a high of $411,469,000 in 1980 to $189,353,000 in 1987. General purpose traineeship grants to schools of public health dropped from $6,842,000 in 1980 to $2,958,000 in 1987. Project grants for graduate training in public health were funded at $4,949,000 in 1980, but dropped to zero funding in 1982 and remained unfunded through 1987. Curriculum development grants, funded at $7,456,000 in 1980, were not funded at all in 1981 and 1982, but then recovered with funding at $1,740,000 in 1983, then at $2,856,000 in 1984 rising to $9,787,000 in 1987. Grants for graduate programs in health administration were funded at $2,967,000 in 1980, dropped to $726,000 in 1981, and then rose to $1,416,000 in 1982 where funding remained fairly steady, with 1987 levels at $1,482,000 (U.S. DHHS, 1988).

Funding has continued to be problematic for public health education programs and schools of public health. Through the 1990s funding levels remained nearly constant. During that time tuition and other costs continued to increase, resulting in a reduction in the amount of public health professional education actually provided. At the beginning of the 21st century, a major barrier to workforce development was found indicating a "incredibly weak" budget allocated for training (Gebbie, 1999; PHLS, 1999).

Following the events of September 11, 2001, there has been new interest in public health and promises of increased funding. If used wisely, these promised funds will strengthen the public health system through investments in both needed technologies and properly educated and prepared public health professionals. The schools of public health were examined in greater detail and the progress made was described in the landmark report, The Future of Public Health (IOM, 1988).

2.7.4 Current Status

Many college graduates who work in public health are educated in other disciplines. For example, of the total public health workforce, nurses make up about 10.9% and physicians comprise about 1.3% (Center for Health Policy, 2000). The Health Resources and Services Administration (HRSA) list of categories of public health occupations includes administrators, professionals, technicians, protective services, paraprofessionals, administrative support, skilled craft workers, and service/maintenance workers. Within these categories fall a number of different kinds of positions including administrative/business professional, public health dental worker, public health veterinarian/animal control specialist, environmental engineering technician, and community outreach/field worker.

Within public health education, the basic public health degree is the M.P.H., while the Doctor of Public Health (Dr.P.H.) is offered for advanced training in public health leadership. There are also individuals working in public health who receive academic degrees (e.g., M.S. and Ph.D.) in public health disciplines such as epidemiology, the biological sciences, biostatistics, environmental health, health services and administration, nutrition, and the social and behavioral sciences. The public health workforce also includes many professionals trained in disciplines such as social work, pharmacy, dentistry, and health and public administration.

Most persons who receive formal education in public health are graduates of one of the 32 accredited schools of public health or of one of the 45 accredited M.P.H. programs. The Council on Education for Public Health (CEPH) is responsible for adopting and applying the criteria that constitute the basis for an accreditation evaluation. In 1998–1999 there were 5,568 graduates from the then 29 accredited schools of public health (ASPH,

2000). The majority of these graduates (61.5%) earned an M.P.H. degree, an additional 28.4% received a master's degree in some other discipline, and 10.1% earned doctoral degrees (ASPH, 2000). According to a survey conducted by Davis and Dandoy (2001), the 45 accredited programs in Community Health and Preventive Medicine (CHPM) and in Community Health Education (CHE) graduate between 700 and 800 master's degree students each year.

The other programs where students receive master's level training in public health are in public administration and affairs, health administration, and M.P.H. programs in schools of medicine. In 1997–1998 an unknown number of the 9,947 graduates of master's degree programs in public administration and affairs (M.P.A.) emphasized public health in their training (NASPAA, 2002). The Association of University Programs in Health Administration report that in 2000 there were 1,778 graduates who received master's degrees, with some (again an unknown number) of them the M.P.H. and M.S. degrees (AUPHA, 2000). In 1998 of the 125 accredited U.S. medical schools, 36 medical schools offered a combined M.D./M.P.H. degree, and 56 reported that they taught separate required courses on such topics as public health, epidemiology, and biostatistics (Anderson, 1999). Public health workers also receive undergraduate training from colleges or universities that offer programs in the environmental sciences or in health education and health promotion.

While it is unclear exactly how many public health workers there are in the United States today, it is estimated that about 450,000 people are employed in salaried positions in public health, and an additional 2,850,000 volunteer their services (Center for Health Policy, 2000). Additionally, the exact number of workforce cannot be determined as limited information is obtained regarding the numbers of volunteers and salaried staff in voluntary agencies. Persons who graduate with training in public health are, however, only a small portion of the public health workforce. Nationally, it has been estimated that 80% of public health workers lack specific public health training (CDC, 2001c) and only 22% of chief executives of local health departments have graduate degrees in public health (Turnock, 2001).

2.7.5 Schools of Public Health

Schools of public health vary in many ways including size, organization, and degrees offered. Most of the schools offer courses in the five areas identified as core to public health: biostatistics, epidemiology, environmental health sciences, health services administration, and social and behavioral sciences. In addition, some schools also offer courses such as nutrition, biomedical, laboratory sciences, disease control, global health, and health policy and management.

2.7.6 Progress in Schools of Public Health

In 1988 the Institute of Medicine (IOM) reports, The Future of Public Health, which describe the field of public health as being in disarray (IOM, 1988). The focus of that report was on public health practice but it did have a number of recommendations for schools of public health.

- New linkages between public health school's programs, and public health agencies at the federal, state, and local levels.

- The development of new training opportunities for professionals who are already practicing in public health.

- Development of new relationships within universities between public health school's programs and other professional department.

- Conducting a wide range of research that includes basic applied research and research in program evaluation and its implementation.

- More extensive approaches to education that encompass the full scope of public health practice.

- Strengthening the knowledge base in the areas of international health and the health of minority groups.

The report also urged public health schools to serve as resources to government at all levels in the development of public health policy.

In summary, the task defined by the IOM report was "to assist the schools in developing a greater emphasis on public health practice and train them with knowledge that matches the scope of public health" (IOM, 1988).

Fineberg and colleagues (1994) identified that the 1988 IOM report states "the professional education should be grounded in the 'real world' public health" as the most influential recommendation. This recommendation generated a number of initiatives that aim to establish a closer relationship between public health schools and public health practitioners. One of the first efforts of IOM report was a collaborative study by the Johns Hopkins School of Hygiene and Public Health and the ASPH (funded by HRSA and CDC in 1989) to define the essential elements of the profession of public health. Public health practitioners and faculty from the public health school were brought together in the Public Health Faculty/Agency Forum, issuing a report in 1991 that emphasized the following:

- Public health education which is based upon universal competencies of public health practice.

- Cooperation between public health schools and public health agencies, including supervised practice for students (Fineberg et al., 1994).

The forum also recommended changing accreditation criteria to emphasize the practice of public health. In response, CEPH revised accreditation criteria to include a required practicum experience.

In 1991 the Council on Linkages Between Academia and Public Health Practice was established to "promote activities that link public health academic programs with the practice community through refining and implementing the forum recommendations" (Eisen et al., 1994). The Council, which includes representatives from national public health academic institutions and practice organizations, has initiated many efforts to enhance academic collaboration. These include demonstration programs that examine academic/practice linkage approaches (Bialek, 2001), a national public health practice research agenda (Conrad, 2000), and a set of core competencies for public health professionals. The core competencies are organized around three job categories—front line staff, senior level staff, and supervisory management staff (Council on Linkages, 2001).

It requires active partnerships between the community and researchers who may or may not be members of that community. Partnerships and coalitions are important in developing prevention and health promotion programs, because no single agency has all the resources, access, and trust relationships to address the wide range of community determinants of public health problems (Green et al., 2001).

Other approaches for strengthening ties between schools of public health and public health practice were reported in a survey of schools of public health. The committee conducted a survey of public health school that listed recommendations from The Future of Public Health (IOM, 1988) and asked schools to indicate what they had done in response. The survey was mailed by ASPH in February 2002 to all accredited schools, out of 31 accredited public health school, 25 responded to the survey which included Boston University, Harvard University, University of California Berkley and many others.

One key recommendation in the 1988 report demonstrates the linkages of state and local health departments, which are important to strengthening ties with the community. Each of the respondent schools indicated that at least some of its faculty have professional working relationships with state or local health departments or both. Their activities include conducting requested research projects, providing technical assistance, serving as the local epidemiologist or health officer, providing staff development or training and serving on professional advisory committees. Major barriers to student involvement in activities with state and local health agencies were identified as lack of financial support and geographical distance from the health department.

The survey also asked about the importance of practice experience as criteria for admission of student applicants and faculty hiring process. For faculty recruitment, prior practice experience was rated very highly important or important by about one-third (32%) of the respondent schools while for student admission about one-half or 52% of schools rated prior experience very important or important.

Ties between public health school and the practice communities have been strengthened, but barriers remain the same. Foremost, among the barriers is a lack of funding and incentives for such activities. As discussed earlier, schools of public health obtain most of their funding primarily through research grants and contracts, because federal support for teaching and practice activities has declined enormously during the past two decades and has not been replaced by state or private sources of funding. Additionally, the incentive and reward structure for faculty tenure and promotion is weighted heavily toward research, publication, teaching and practice activities.

Another 1988 recommendation for linking schools to practice is for schools to participate in policy development. The survey asked schools to indicate how they fulfill their potential role as significant resources to government at all levels in the development of public health policy as well as barriers to engaging in this role. The vast majority of schools that responded to have faculty who engage in numerous policy development activities such as:

1. Policy development for legislative bodies (23 schools responded positively),

2. Public health advocacy with State Government (23 schools responded positively),

3. Public health advocacy with local Government (22 schools responded positively),

4. Research requested by state policy makers (23 schools responded positively),

5. Research requested by local policy makers (21 schools responded positively) and

6. Public health workforce development (24 schools responded positively).

(Information based on https://www.ncbi.nlm.nih.gov/books/NBK221176/)

New training opportunities. The Future of Public Health (IOM, 1988) recommended that schools of public health improve their educational approaches for the practicing public health workforce through short courses and continuing education. Currently, all accredited schools of public health offer continuing education for public health professionals, as do the accredited programs. The overarching goal of continuing professional education is to educate and support public health professionals through enhancement of their knowledge and skills in public health practice, theory, research, and policy. Continuing education is an essential component of any career (Gordon and McFarlane, 1996), and all schools and practice agencies should develop appropriate support systems for relevant continuing education for public health practitioners.

One approach to continuing education is to offer yearly conferences or workshops on specific topics. These programs can be sponsored by universities or in partnership with public health programs, agencies, or associations. Certification programs are another approach to educate those people currently working in public health. About one-third of the accredited schools of public health currently offer certification programs. Standards for admission and completion vary across the schools. Certification programs emphasize core courses of public health concepts from the five core content areas taught in M.P.H. programs, that is, epidemiology, biostatistics, environmental health sciences, health services administration, and social and behavioral sciences. Others focus on a specific content area such as international health, environmental health, occupational health, injury control, health policy, and health administration.

The CDC Graduate Certification Program (GCP)—a program no longer funded—was a prime example of certification programs. It was designed for CDC field officers, state health department personnel, and selected others with at least three to five years of experience in public health practice. The program allowed CDC Public Health Advisors working in state and local health departments to earn a graduate certificate in public health and was available from one of four accredited schools of public health which are: Tulane University School of Public Health and Tropical Medicine, Emory University Rollins School of Public Health, Johns Hopkins University Bloomberg School of Public Health, and University of Washington School of Public Health and Community Medicine.

Academic institutions (including schools of public health) also offer summer courses. Subjects range from basic biostatistics, epidemiology, and Geographic Information Systems (GIS) applications, for management and administration. Such programs can vary in length from a single one-day course to multiple week-long courses. Another approach to traditional continuing education programs, as described by Halverson and colleagues (1997), involves the creation of master's- and doctoral-level executive programs that minimize time lost from work through use of distance learning teaching methods. By enabling workers to continue in their work responsibilities while completing would be helpful to reduce burden.

The introduction of Web-based tools for education is producing a major change in the way schools and colleges conduct classes, particularly in the area of continuing education. The use of such technology is referred to as distance learning (Riegelman and Persily, 2001). This development builds upon more than two decades of computer networking activities (email and bulletin board systems), and the increased availability of the Internet has produced phenomenal growth in the extent and scope of online education. Distance learning today has become an important alternative to traditional methods of education, because the existing technology is very efficient in facilitating complicated distance learning environments and highly structured learning methods (Mattheos et al., 2001). The Public Health Training Network (PHTN) is an example of successful distance learning program. This network has been linked to nearly one million people to train them in a variety of formats: print-based self-instruction, interactive multimedia, videotapes, two-way audio conferences, and interactive satellite videoconferences (CDC, 2001b).

Links with other departments and schools. The Future of Public Health (IOM, 1988) recommended that schools of public health develop new relationships with other schools and departments both within their universities as well as with other institutions. According to survey data. For example, 96% of reporting schools (n = 24) indicated that their public health students could take courses in schools of medicine that would count toward their degree, as did 64% (n = 16) for courses in nursing, 44% (n = 11) in dentistry, 68% (n = 17) in law, and 72% (n = 18) in social work. 56% (n = 14) schools reported that students "often" avail themselves of these opportunities in other schools and departments. Whereas, only 28% responded "sometimes" as their availability.

Research in schools of public health should range from basic research in fields related to public health, through applied research and development, to program evaluation and implementation research (IOM, 1988). To describe the range of research conducted in schools of public health, the committee survey asked each school to estimate the percentage of research undertaken at their school, and it would be characterized by the following:

- Basic or fundamental research: Research conducted for the purpose of advancing our knowledge;

- Applied research: Research designed to use the results of other research (e.g., basic research) to solve real world problems;

- Translational research: Research on approaches for translating results of other types of research to community use; or

- Evaluative research: The use of scientific methods to assess the effectiveness of a program or initiative.

Among respondent schools, the distribution of the types of research undertaken varied greatly. On average, applied research was reported most often (35% mean, range of 10%–60%), followed by basic research (27% mean, range of 0%–70%), evaluative research (20% mean, range of 1%–50%), and translational research (17% mean, range 0%–30%).

For broadening the scope of public health education, the 1988 IOM report recommended that schools of public health provide an opportunity to learn the entire range of skills and knowledge necessary for public health practice. Recent efforts to encompass a broad scope of education have focused on identifying basic competencies in public health and on developing curriculum, which teach us the information and skills necessary to meet those competencies. The CDC Office of Workforce Policy and Planning (CDC, 2001c) has developed a table of public health competency sets. One of these is a set of core competencies developed by the Council on Linkages Between Academia and Public Health Practice (Council on Linkages, 2001). The ASPH also endorsed the Council on Linkages competencies and plans to develop complementary competencies for M.P.H students. The committee survey of the public health school requested respondents to indicate courses that they offer students in cultural or international health as well as other selected areas: (based on the data reported in https://www.ncbi.nlm.nih.gov/books/NBK221176/).

1. Cultural Competencies: 16 schools responded positively out of 25 responded

2. Ethics: 22 schools responded positively out of 25 responded

3. Health Disparities: 19 schools responded positively out of 25 responded

4. Social Justice: 17 schools responded positively out of 25 responded

5. Human Rights: 13 schools responded positively out of 25 responded

6. International Global Health: 18 schools responded positively out of 25 responded

7. Social Epidemiology: 15 schools responded positively out of 25 responded

The final question on the committee survey of public health school asked for the input on identifying the most important challenges and opportunities faced by public health school and M.P.H. programs over the next 10 years.

Survey responses identified challenges and opportunities. According to respondents, public health as a profession is not well defined, lack of clear definition is one reason the public does not understand the field. Raising public awareness of public health's contributions to health and quality of life is important. Such awareness would help assure adequate support for public health programs. Lack of support and funding was a major issue identified frequently. Respondents indicated that increased funding is needed to support students and workforce development, and is critical to maintaining stable support for key academic programs including teaching.

Respondents indicated that the changing environment and ever-widening scope of public health requires collaboration and partnerships with other disciplines. Additionally, within the field, schools need to build strong relationships among academia, scientists, and the professional practice community, thereby to get benefited from each other.

Education and training issues were identified by numerous responders. One person wrote, "Public health is no different than other academic programs in that we tend to produce graduates for yesterday's workplace and yesterday's problems. Producing M.P.H graduates responsive to what is needed today requires an understanding of the driving forces that affect public health practice and the public health workforce." Respondents indicated that major needs include understanding that multiple factors influence health and that public health issues require societal change as well as changes in individual behavior for risk reduction. One respondent indicated that the primary goal of schools of public health should be to train the next generation of leaders as public health scientists and public health professionals, stating that "Research informs practice and policy. Leadership guides them all." Other educational or training issues included:

- Education in the M.P.H. level should be comprehensive, integrated, and broad-based to support the need for general public health and emergency preparedness.

- M.P.H. programs need to be redesigned to permit greater flexibility in the development of clusters of skills and competencies in response to the rapidly changing public health environment.

- Baccalaureate training in schools would provide a vehicle for attracting a new cadre of students into public health.

- There is a need for opportunities for training in non-degree programs for part-time and mid-career students, and for increased distance learning programs.

- There is a need for more practical experience for graduates.

Issues addressed were that there was a need to recruit minority faculty to achieve diversity in faculty hired but it was difficult to hire in specific disciplines such as biostatistics and epidemiology, and that it is necessary to maintain and improve faculty salary levels to be competitive with other sectors. Another issue identified as important was building the public health infrastructure. Some respondents indicated that there should be national attention and standards for trained personnel, along with funding to meet those standards. Respondents indicated that schools should be expected to be a resource to provide training and to meet these standards and that a lack of standards and funding results in an inadequately prepared public health workforce. It was suggested that certification or credentialing of public health professionals is an important issue. It was proposed that schools assist in the accreditation process for local departments of health by helping them meet their continuing education needs.

Respondents also indicated that the emphasis of public health research must be reviewed periodically. Schools of public health must more effectively promote prevention as a powerful means of health protection. Public health must find new approaches to reach the public on a level that effectively encourages primary prevention and enables individuals to change known risk behaviors to healthy behaviors. There should be increased emphasis on partnerships to develop viable research programs. Understanding and addressing the determinants of ethnic and racial health disparities, and bioterrorism had to be the focus during research. Finally, respondents identified, but did not elaborate on, the following challenges:

- Globalization

- Re-emerging infections

- Human genome

- Quality of healthcare

- Uninsured and underinsured populations

- Population aging

2.8 SUMMARY

The establishment of the Johns Hopkins University School of Hygiene and Public Health in 1918 marked the beginning of public health education. There are currently 32 accredited schools of public health and 45 accredited community health programs. The Council on Education for Public Health estimates that the total number of accredited schools and programs may well double within the next 10 years and that the most dramatic growth is occurring outside the established schools of public health. Many of the nation's accredited medical schools now have operational M.P.H. programs or are currently developing a graduate public health degree program (Evans, 2002). New specializations are emerging such as human genetics, management of clinical trials, and public health informatics. Many schools and competing organizations are involved in distance learning programs that offer the possibility of fulfilling the long-recognized need to bring public health education to the homes and offices of the public health workforce. The Internet also offers the possibility of bringing public health education to populations across the country and around the world; indeed, health information sites are among the most popular and frequently visited of all Web applications.

Previous efforts to design truly effective systems of public health education generally floundered because of a lack of political will, public disinterest, or a paucity of funds. Since September 11, 2001, however, the context has changed dramatically. With public health rising high on the national agenda and an abundance of funds being promised, perhaps there is now an opportunity to shape a future system of public health education that addresses the problems that have been so often described and analyzed.

CHAPTER 3

Formal and Informal Learning in Continuing Professional Education in Public Health

3.1 BACKGROUND OF CONTINUING PROFESSIONAL EDUCATION

Varying demographic backgrounds, including education and discipline of public health educators, warrant some basic standards of practice. As cited in Kasworm, Rose, and Ross-Gordon (2010), typical features of a profession are proficiency, performance, ethical standards (with sanctions). These typical features of a profession need to be maintained and improved through a learning process.

Houle (1980), as cited in Kasworm et al. (2010), identified 14 features or goals of professionalization in regards to the entry into and continuing education in the health professions. The features include agreement on the defining function and mission of the profession, mastery of theoretical knowledge, capacity to solve problems, use of practical knowledge, self-enhancement, formal training, credentialing, creation of subculture or community of practice, legal reinforcement, public acceptance, ethical practice, penalties, relation to other vocations, and relation to users of service. Public health professionals need to maintain their integrity as professionals by following the ethical standards and providing the best quality of service to the community.

A study by Glascoff, Johnson, Glascoff, Lovelace, and Bibeau (2005) showed that most public health educators in North Carolina are white females; most do not have a Certified Health Education Specialist (CHES) designation; younger health educators are more likely to have health education degrees; and almost two-thirds of public health educators have administrative responsibilities. Another study by Finocchio, Love, and Sanchez (2003) showed that in the San Francisco Bay Area, there were 4 health educators with a Master of Public Health (M.P.H.) per 100,000 persons in 1999, and the majority worked in local health departments and community-based organizations. The different needs of public health educators related to their competencies create challenges for the educational providers to include them all in one single continuing professional education (CPE) activity. Furthermore, the Centers for Disease Control and Prevention (CDC) proposed genomic competencies to be added to the current competencies of public health educators. The need to expand the expected skills of public health educators was revealed in a study by Chen and Goodson (2007). Although the participants in the study have negative attitudes, low awareness, and deficient genomic knowledge, the study shows that expected skill competencies of public health educators have expanded and corresponding curriculum adjustments are needed (Chen and Goodson, 2007).

There are many opportunities for public health professionals to engage in learning in the profession. These learning opportunities may be conducted formally by a particular organization, but some may be experienced as informal or incidental learning in the workplace. According to Marsick and Watkins (1990), formal, informal and incidental learning are distinguished based on the degree of control by the learner. Formal learning is typically highly structured, classroom-based, and organized by a particular organization. Informal learning can be encouraged by organization but mostly occurs in daily life where individuals learn from and through experience in a particular situation. Incidental learning is a sub-category of informal learning that almost always takes place in everyday experience as a byproduct of another activity.

Formal learning opportunities are commonly organized by organizations or educational providers in the form of education and training to enhance specific skills of public health professionals. These formal learning opportunities may be designed as conferences and seminars organized by associations of public health professionals. One of the biggest conferences for public health professionals is the one organized by the American Public Health Association (APHA). The APHA conference is held annually in different cities across the United States. There are annual meetings and conferences for the public health field that are also available locally. In the state of Georgia, the Georgia Public Health Association (GPHA) and the Georgia Rural Health Association (GRHA) held the annual meetings and conferences for public health professionals. Some of these seminars or conferences may also be targeted to a specific field in public health. For instance, the Society of Public Health Educators (SOPHE) facilitates up-to-date information only for those professionals who are interested in the public health education field. The seminars, conferences, education and training that are organized by educational providers or professional associations of public health professionals are commonly identified as the formal type of CPE activities.

CPE is needed to help public health professionals stay up-to-date with current development in public health, and maintain or improve mastery in job specific area of competencies. Therefore, in order to provide the best services, public health professionals need to participate in continuing education (CE) on a regular basis. As cited in Abrahamson (1985), "A generally recognized purpose of continuing professional education (CPE) in the health professional is to impact on health practitioners' knowledge and attitudes, their skills in providing health care, and on health-care outcomes of patients" (p. 1). In addition to that, a study by Price, Akpanudo, Dake, and Telljohann (2004) reveals that CPE can fill the gaps in formal preparation during undergraduate or graduate education, provide opportunities to update current skills, and provide new skills needed for current jobs.

The formal, informal and incidental learning transform the knowledge, beliefs and behaviors of public health professionals in their professional practices. Through the interactions in the workplace, these changes of knowledge, beliefs and behaviors affect the learning at group or team level that eventually influences the learning at the organizational level. The organization that learns continuously and transforms itself is defined as a learning organization (Watkins and Marsick, 1993). Formal, informal and incidental learning are differentiated by how much control is exercised by the learners (Marsick and Watkins, 1990). Formal learning is commonly sponsored by the institution, conducted in a classroom-based and highly structured setting (Marsick and Watkins, 1990). Similar to formal learning, informal learning may also occur in the workplace, but informal learning must take place with the collaboration of others where individuals consciously learn in the workplace in a non-routine situation (Marsick and Watkins, 1990). According to Marsick and Watkins (1990), professionals may learn informally from and through experience when they make sense of situations that they encounter in their daily lives. A subset of informal learning is called incidental learning (Marsick and Watkins, 1990). Marsick and Watkins (1990) described the difference between informal and incidental learning. According to Marsick and Watkins (1990) "Informal learning: is predominantly experiential and non-institutional... [and] incidental learning: is unintentional, a byproduct of another activity" (p. 7). In addition, Marsick and Watkins (1990) further described that in informal learning: (1) the learners have more control of their learning; (2) it does not have to be delivered in a classroom; and (3) the outcomes are less predictable than formal learning. As for incidental learning, the knowledge is "usually tacit, taken for granted and implicit in assumptions and actions" (Marsick and Watkins, 1990, p. 7).

3.2 FORMAL LEARNING IN CONTINUING PROFESSIONAL EDUCATION

In regard to formal learning, previous studies showed evidence of the advantages for professionals in participating in CPE. First, attending CPE on a regular basis fills the knowledge gaps that public health professionals experience due to their different backgrounds of knowledge. Public health professionals come from a variety of backgrounds and work in a variety of settings with the common goal of promoting population health (Gebbie et al., 2003b). Public health professionals may come from medical, sociology, anthropology, nursing, and other areas. CPE activities aim in bridging gaps between their prior knowledge and the knowledge needed in their professional work in solving current public health problems.

A study by Price et al. (2004) revealed that CPE can fill the gaps in formal preparation during undergraduate or graduate education, provide opportunities to update current skills, and provide new skills that are needed for their current jobs. However, these needs of knowledge and skills are varied and job-specific to the need of public health professionals. Therefore, it is necessary that the instructor be able to facilitate the diverse need of CPE participants. A study by Ellery, Allegrante, Moon, Auld, and Gebbie (2002) revealed that CPE programs have been shown to be most effective when tailored to the specific need of the participants.

Second, CPE is offered by multiple providers and can be developed in many formats. This flexibility of CPE can be tailored to facilitate the improvement of education and knowledge for health professionals. CPE can be designed by any educational providers, universities, and professional associations to stay up-to-date with the latest research or issues in the public health field. This advantage is supported by Desikan (2009) who defined CPE as "any educational activity, formal or informal, that professionals undertake to help them understand their profession and perform better at their work" (p. 1). The form of CPE activities may include but are not limited to seminars, conferences, workshops, academic courses, satellite instruction, training programs, and directed self-study programs. CPE with adequate content, experienced instructors, and effective methods of delivery support the professional development of the public health professionals to better human life and service in a profession (Desikan, 2009).

Third, CPE may offer credits that can be used to maintain professional certification of public health professionals. Many sessions in the APHA conference offer Continuing Education Contact Hours (CECH) for the participants. For instance, CECH were offered in seminars or conferences as a requirement for public health educators to maintain professional certification. Additional fees are required in order to attend these sessions. Each session offers 1–5 credits that can be submitted to the National Commission for Health Education Credentialing (NCHEC) to maintain professional certification. These contact hour credits can be obtained through various CPE activities that are acknowledge by NCHEC. NCHEC certifies public health educators and develops standards to maintain this professional certification through competency-based examination and CPE activities (NCHEC, 2013).

NCHEC requires public health educators who hold Certified Health Education Specialist (CHES) and Master Certified Health Education Specialist (MCHES) to accumulate a minimum of 75 CECH over the five-year certification period (NCHEC, 2013). This means that CHES are encouraged to accumulate a minimum of 15 CECH per year and to complete all continuing education requirements at least 90 days prior to recertification (NCHEC, 2013).

Although NCHEC does not specify the type of CPE activity that public health educators should attend, it is recommended that CHES and MCHES choose the CPE activities that can enhance their knowledge in their areas of responsibility. These areas of responsibility are (1) Area I: Assess Needs, Assets and Capacity for Health Education; (2) Area II: Plan Health Education; (3) Area III: Implement Health Education; (4) Area IV: Conduct Evaluation and Research Related to Health Education; (5) Area V: Administer and Manage Health Education; (6) Area VI: Serve as a Health Education Resource Person; and (7) Area VII: Communicate and Advocate for Health and Health Education (NCHEC, 2013). All of these responsibilities need to be possessed by public health educators in order to provide the best quality of service in their professional work. Although CPE can come in many formats, a study by Davidson (2008) found that public health educators preferred to (1) attend seminars or conferences; (2) attend professional associations' annual meetings; and (3) complete home self-study print materials in order to continue their professional learning.

The fourth advantage of CPE is to build a network with other professionals with similar interests. This peer networking provides a sense of professionalism for public health professionals in terms of proficiency, performance, and ethical standards in their work which are the typical features of a profession, according to Kasworm, Rose, and Ross-Gordon (2010). These typical features of a profession need to be maintained and improved through CPE activities. This type of activity is supported by a study by Hirotsugu (2006) that revealed the top reasons for participation among health workers in Ghana were to maintain and improve professional knowledge and skills, to interact and exchange views with colleagues, and to obtain a higher job status.

The formats of CPE are mostly seminars or conferences that are organized by professional associations or educational providers. Despite all the benefits that CPE offers, there are several problems that will hinder the achievement of improving knowledge and skills of the participants. Some problems that will be discussed in this section are related to various resource constraints and needs of public health professionals that lead to low participation and low quality of learning in CPE activities.

All of these concerns need to be addressed in order to maintain high participation in CPE programs and improve the job performance of public health educators.

Ideally, professionals will optimize their learning if CPE can cover all three aspects of learning as Houle (1980) described in his study. Houle (1980) explained how professionals learn (1) through instruction in which the educators decide what professionals are required to know; (2) through inquiry in which professionals express and learn new techniques or concepts using cooperation methods; and (3) through performance in which professionals learn through practice in the actual work settings. Public health professionals may benefit more by learning in the workplace by finding solutions for the problems at hand rather than participating in a limited time of CPE activities.

3.3 INFORMAL LEARNING IN CONTINUING PROFESSIONAL EDUCATION

Public health professionals learn informally in the field through multiple activities by being a member of groups or teams that serve different purposes. According to Watkins and Marsick (1993, p. 69) informal continuous learning of professionals can occur because of the following:

1. Unanticipated experiences and encounters, the learning may or may not be consciously recognized or acknowledged by learners;

2. New job assignments and participation in teams;

3. Self-initiated and self-planned experiences through the use of media, mentor, conference, travel or consulting;

4. Total quality groups designed to promote continuous learning;

5. Planning a framework for learning through career plans, training, or performance evaluation;

6. Combination of less organized experiences; and

7. Just-in-time courses.

Through these activities, public health professionals are continuously learning informally in the field. As cited in Kasworm, Rose, and Ross-Gordon (2010), typical features of a profession are proficiency, performance, and ethical standards (with sanctions). These typical features of a profession need to be maintained and improved through a learning process. Houle (1980) as cited in Kasworm, Rose and Ross-Gordon (2010), identified 14 features or goals of professionalization in regards to the entry and continuing learning in the professions. The features are agreement on the defining function and mission of the profession, mastery of theoretical knowledge, capacity to solve problems, use of practical knowledge, self-enhancement, formal training, credentialing, creation of subculture or community of practice, legal reinforcement, public acceptance, ethical practice, penalties, relation to other vocations, and relation to users of service (Kasworm, Rose and Ross-Gordon, 2010). Public health professionals need to maintain their integrity as professionals by following the ethical standard and providing the best quality of service to the community.

Public health professionals experience informal or incidental learning in the workplace during their daily practice from within or outside the organizations. According to Marsick and Watkins (1990), informal learning is "...predominantly experiential and non-institutional... [and takes place in]... the normal course of daily events without a high degree of design or structure" (pp. 7 and 14). Public health professionals learn and share knowledge among each other so they can improve their job performance (Pereles, Lockyer, and Fidler, 2002). A study by Pereles, Lockyer, and Fidler (2002) also found that within this informal learning community, members appeared to be supportive to enhance each other's learning. They also found that informal learning also encourages members to freely give opinions to agree or disagree rather than focuses on the "right" answer or achieve a consensus.

Professional learning is central to informal learning because professionals have motivation to learn and possess the ability to develop strategy to pursue their learning needs (Watkins and Marsick, 1993). Many studies showed that the interaction between health professionals in the workplace is able to foster learning. This informal learning can be found around real-life experience of public health professionals. One of the well-known informal learning groups that facilitate informal and incidental learning for public health professionals to complement their formal learning is known as Communities of Practice first introduced by Wenger (Wenger, 1998).

Community of Practice gains its popularity especially in public health field. As mentioned earlier, the public health area consists of a variety of professional backgrounds working together toward a common goal. This realization leads to the interest in Community of Practice or a related notion that a public health professional may belong to more than one community. This notion could be an important source of enrichment through exchanging ideas, points of view, questions, and practices from a variety of Communities of Practice (Falk and Drayton, 2009). According to Wenger, McDermott, and Snyder (2002), Communities of Practice are "groups of people who share a concern, a set of problems, or a passion about a topic, and who deepen their knowledge and expertise in this area by interacting on an ongoing basis" (p. 4).

This approach suggests that participating in a Community of Practice characterizes a pathway for continued professional learning in a community of people who shares the same practice (Falk and Drayton, 2009). As cited in Ho et al. (2010), There are five unique phases in the evolution of Community of Practice: potential, coalescing, maturing, stewardship, and transformation. Ho et al. (2010) continued that there are four fundamental characteristics of Community of Practice in the health sector: active social interaction, knowledge sharing, knowledge creating, and identity building. Small or big Communities of Practice are determined by the number of members who could influence the structure of Communities of Practice whether divided by geographic regions or by sub-topic to encourage active participation from members (Wenger, McDermott and Snyder, 2002). Community of Practice is also different in terms of its life span as some Communities of Practice exist for centuries and others become inactive after a short period of time (Wenger, McDermott and Snyder, 2002). Communities of artisans, such as the communities of violin makers, usually exist for many centuries due to the inherited knowledge from generation to generation within the Communities of Practice (Wenger, McDermott and Snyder, 2002).

The advancement in technologies contributes to the collocated or distributed forms of Communities of Practice. Wenger, McDermott and Snyder (2002) describe that although many Communities of Practice started among workers at the same organization or who live nearby, many Communities of Practice also exist across geographical and professional boundaries. Such distribution can influence the type of interaction between members. Some members may meet every day, but others may only interact through email and telephone and only meet once or twice per year. The existence of a shared practice (not the type of interaction) allows the member of Communities of Practice to share useful knowledge. Wenger, McDermott and Snyder (2002) explain that the advancement in technology and the need of globalization might make the distributed Communities of Practice as a standard rather than an exception.

Communities of Practice also can vary according to the background of their members. According to Wenger, McDermott and Snyder (2002), homogeneous Communities of Practice consist of people with similar backgrounds or functions. This type can be advantageous in the forming phase of the Communities of Practice, but may have the disadvantage of having limited perspective in solving the complex problems. A shared practice with members from various backgrounds (e.g., a heterogeneous Community of

Practice) can contribute to the solution of these complex problems from multiple perspectives.

The topic discussed by members can also influence the form of Communities of Practice. Wenger, McDermott and Snyder (2002) explained that Communities of Practice can have members within the same business (inside boundaries) and members from across business units (across boundaries). Within the same business, Communities of Practice are created as people prefer to address recurring problems in their job with their peers through Communities of Practice rather than memorize everything independently. Across businesses, Communities of Practice are created by peers in various departments in the same company to solve problems and develop common guidelines, tools, standards, procedures, and documents. Communities of Practice can also be created across organization boundaries where members interact to learn new knowledge outside their company affiliation and job description.

The two forms of Communities of Practice discussed by Wenger, McDermott and Snyder (2002) are spontaneous or intentional, and unrecognized to institutionalized Communities of Practice. Communities of Practice can be created because members need each other as learning partners without intervention from their organization or can be created intentionally as the means to administer a skill that is needed by an organization (Wenger, McDermott and Snyder, 2002). Therefore, Communities of Practice are also unrecognized by the organization and only take form as informal lunch discussions or can be institutionalized as the official structure of the organization, and when well managed can offer legitimacy and useful resources for members.

Although Communities of Practice can take many forms, they all share three fundamental elements: (1) a domain of knowledge, (2) a community of people, and (3) a shared practice (Wenger, McDermott and Snyder, 2002). The three elements can make Communities of Practice an ideal knowledge structure when they function well together (Wenger, McDermott and Snyder, 2002). Wenger, McDermott and Snyder (2002) defined knowledge structure as "a social structure that can assume responsibility for developing and sharing knowledge" (p. 29). This model, according to them, can provide a common language that facilitates discussion, collective action, efforts to gain legitimacy, sponsorship, and funding in an organization.

According to Wenger and coworkers, the domain of knowledge defines a set of matters in Communities of Practice as common ground that can create a sense of common identity. They described that domain of knowledge validates the purpose and value of the members. The members are encouraged to contribute and participate in the discussion and to present their ideas. Participating in Communities of Practice helps members to guide their learning and gives meaning to their action, to decide which information is worth sharing, to pursue a certain activity, and to recognize the potential in tentative or half-baked ideas.

According to Wenger and coworkers, Communities of Practice can foster interaction and relationships based on mutual respect and trust so that members can share ideas, expose one's ignorance, ask difficult questions, and listen to the opinion of other members in Communities of Practice. The shared practice aspect of Communities of Practice is "a set of framework, ideas, tools, information, styles, language, stories, and documents that community members share" (Wenger, McDermott and Snyder, 2002, p. 29). They further stated that through the interaction in communities of practice, members develop expertise of the basic knowledge that enables the community to proceed efficiently in dealing within its domain.

According to Wenger, McDermott and Snyder (2002, p. 51), seven principles cultivate Communities of Practice:

1. Design for evolution.

2. Open a dialogue between inside and outside perspective.

3. Invite different levels of participation.

4. Develop both public and private community spaces.

5. Focus on value.

6. Combine familiarity and excitement.

7. Create a rhythm for the community.

Wenger and coworkers also describe the five stages of developing Communities of Practice: (1) potential, (2) coalescing, (3) maturing, (4) stewardship, and (5) transformation. The potential and coalescing occur during early stages when Communities of Practice are in the planning and launching phase, the rest occur in the advanced stages when Communities of Practice are in the growing and sustaining phase. In the potential stage, the members are still defining the Community of Practice of domain; finding people who are interested in the domain; and defining the common knowledge in Communities of Practice. In the coalescing stage, members establish values of sharing knowledge, develop relationships with sufficient trust to discuss issues, and discover what and how to share the specific knowledge within the communities of practice. In the maturing stage, members define the role of the Community of Practice within an organization and its relationship with other domains, manage the boundaries of the communities of practice, and seriously organize and administer relevant knowledge. The core members identify the gaps in the community's knowledge, identify its cutting edge, and feel a need to be more systematic in their definition of the community's core practices. In the stewardship stage, members maintain the relevance of the domain and find a voice in the organization, maintain active engagement in the intellectual focus of the Communities of Practice, and keep them on the cutting edge.

The transformation stage according to Wenger, McDermott and Snyder (2002) occurs when Communities of Practice simply fade away and lose members and energy until no one appears at a community's events or communicates through electronic Communities of Practice. The transformation can also be because Communities of Practice are turning into a social club where the members' focus slowly shifted from core issues to organizational ones, and then to their personal lives, or when Communities of Practice are split into distinct communities or merge with others because of overlapped topic. Finally, according to Wenger and coworkers, Communities of Practice can become institutionalized as a center of excellence or become actual departments within an organization.

Houle (1980) explained in a study by Desikan (2009) that there are three aspects of learning that were experienced by professionals through Communities of Practice. In the study, Desikan (2009) found that Communities of Practice (1) can help professionals to test new ideas and take risks individually or collectively in their domain; (2) can improve the process of distributed cognition and continuous learning through collective learning; (3) focus on professional practice; and (4) help one to learn what and how to perform, to be competent in their professional work.

Communities of Practice can also specifically target the professional practice of public health professionals in their workplace through dealing with specific problems that reside beyond CPE activities. Through Communities of Practice, public health professionals learn how to deal with public health issues in their own professional boundaries. This on-the-job learning is supported by Wenger (1998) who stated that, "learning involves an interaction between competence and experience...[requiring] a constant fine tuning between [the two]" (p. 215). Communities of Practice enable professionals to learn in their own workplace. Such learning helps public health professionals to test new ideas and distributed cognition and continuous learning after the CPE program has concluded. Desikan (2009) stated that, "education is demonstratively more effective when it seeks to improve the ways that professionals actually reason and make decision in their daily practice" (p. 17). This demonstration of effective education can be found in a Community of Practice through its ability to provide immediate access for professionals in making an informed decision on their job.

Public health professionals can construct their own meaning from their background and past experiences to become engaged with other members in Communities of Practice. This construction of meaning is the application of a constructivist approach. Constructivists believe that past experience and background influence the level of understanding as individuals construct new knowledge through the interaction with others (Ültanir, 2012). This meaning construction transforms public health professionals' prior beliefs into new knowledge through reflecting on the current situation from engaging in Communities of Practice. The constructivist approach of Communities of Practice enables public health professionals to appropriately adjust their learning to their own unique community. Therefore, learning through Communities of Practice enables public health professionals to continue learning in their profession beyond the CPE.

Through Communities of Practice, public health professionals learn by learning with others more than from others. They provide an opportunity for public health professionals to review their own professional practice. Communities of Practice are "purposeful and strengthen a group's ability to learn from and apply wisdom to everyday life situations" (Stein and Imel, 2002, p. 94). Communities of Practice can support the CPE activities that are conducted for public health professionals by maintaining and enhancing knowledge and skills before and after the CPE programs have ended, especially when CPE providers are still trying to facilitate all the different needs of various professionals through many different purposes, forms, and locations of CPE delivery (Cervero, 1988). Communities of Practice can help bridge these gaps of knowledge and skills that CPE is unable to provide for public health professionals through the three aspects of learning in Communities of Practice.

In order to understand the learning processes within Communities of Practice, we need to understand the nature of knowledge itself. According to Wenger, McDermott and Snyder (2002): (1) knowledge lives in the human act of knowing;

(2) knowledge is tacit as well as explicit; (3) knowledge is social as well as individual; and (4) knowledge is dynamic. In relation to knowledge lives in the human act of knowing, they explained how knowledge is a living process and not a static body of information. Professionals need opportunities to interact with others who have similar challenges with their professions in order to acquire expert knowledge about the problems that they face. As cited in their book, this knowledge of experts is "an accumulation of experience—a kind of 'residue' of their actions, thinking, and conversations—that remains a dynamic part of their ongoing experience" (p. 9). They add that Communities of Practice make knowledge an integral part of their activities and interactions and also serve as a living repository for that knowledge.

In regard to knowledge being tacit as well as explicit, Wenger and coworkers explain how the tacit aspect of knowledge is often the most valuable for professionals, especially from the business standpoint. People are aware that they know more than they can explain, and this knowledge sometimes can be difficult to be present in documents or through tools. Because of the difficulty in replicating tacit knowledge, professionals need to interact in an informal learning such as storytelling, conversation, coaching, and apprenticeship that Communities of Practice can provide. Communities of Practice are able to produce useful documentation, tools, and procedures that are needed by professionals and cover tacit and explicit aspects of knowledge.

Wenger and coworkers also explain that while knowing is individual, knowledge is social as well as personal. A body of knowledge is developed through a series of disagreements including the controversies that can occur between professionals. According to Wenger and coworkers, members participate in the process of producing scientific knowledge through Communities of Practice. This collective nature of knowledge is especially important in a field when the changes occur too rapidly for an individual to master them. Professionals need feedback from their peers to complement and develop their expertise to solve today's complex problems. Communities of Practice welcome strong personalities and encourage debates and controversy as elements of what makes a community vital, effective, and productive.

According to Wenger and coworkers, knowledge is dynamic and continually in motion, as every field is changing at an accelerated rate. New data, inventions, and problems require professionals who can focus on creative problem solving on more advanced issues. They state that because Communities of Practice provide baseline knowledge and common standards that can be understood well by all its members, Communities of Practice members can focus their energy on advancing new knowledge. Communities of Practice can facilitate interaction to help members manage information overload, get knowledgeable feedback on new ideas, and maintain understanding of current thoughts, techniques, and tools.

Wenger and coworkers also explain that Communities of Practice are ideal for administering knowledge by giving practitioners the freedom to acquire the knowledge they need and share this knowledge with other members of the community. Although Communities of Practice can be part of an organization, they can also flourish independently depending on the level of participation of their members and the emergence of internal leadership within them. These factors influence the many forms that these communities can have.

Wenger and coworkers describe how Communities of Practice can take many forms. They categorized them in seven different formats. Communities of Practice can be (1) small or big, (2) long-lived or short-lived, (3) collocated or distributed, (4) homogeneous or heterogeneous, (5) inside or across boundaries, (6) spontaneous or intentional, and (7) unrecognized or institutionalized.

Despite the advantages offered in face-to-face Communities of Practice, active interaction could be a challenge for public health professionals' practices due to the time constraint and heavy workload of the public health field. In order to deal with this challenge, many Communities of Practice were delivered through the use of technologies that cover wide geographical areas. This type is called Electronic Communities of Practice, or some prefer to call them Virtual Communities of Practice. Electronic Communities of Practice may enhance and be enhanced by existing patterns from offline relationships between member, may overcome the barrier of time and geographical challenges, and provide just-in-time access to solve work-related problems for professionals.

Over the past 40 years, Electronic Communities of Practice have gone through rapid and unpredictable changes. In 1971 email became available using the "@" sign and was followed by the use of UseNet as a network-wide discussion board in 1979 (Falk and Drayton, 2009). In 1985, whole earth electronic link (WELL) was created to stimulate formulation of virtual communities to improve teaching (Falk and Drayton, 2009). PLATO, the online computer-assisted instructional system, was added to the various online communities in the 1970s (Falk and Drayton, 2009).

Public health agencies have supported many educational efforts for workers in federally funded public health program areas, and many of these efforts are available through distance technology. Some of these trainings carry continuing credits that are necessary for the public health field. It is critical that these workers attend the CE in their own workforce to ensure that they keep abreast of evolving organizational, ethical, and communication concerns of their Communities of Practice (Gebbie et al., 2003a).

Many designers, as cited in Falk and Drayton (2009), have experimented with online communities for professional development and implemented concepts, such as Electronic Communities of Practice based on past practices. They stated that, "theoretically informed electronic communities can support professional learning as well by enabling the identification and exploration of areas of professional knowledge, making them accessible to reflection and change" (p. 4). Electronic Communities of Practice are a critical part of one's work and learning that is mediated by electronic tools and resources. They also stated that, "the nature of an electronic community is a blend of vision and experience, of design and emergent" (p. 11). Through Electronic

Communities of Practice, professionals can acquire knowledge which they can later apply to their practices, through discussion and exchange of ideas (Ho et al., 2010).

Online formats can present an effective learning process to professionals. As cited by Harris (2009), the U.S. Department of Education noted ample evidence that on-line education is often more effective than face-to-face education, quite possibly because online education participants tend to spend more time on task. Still, the same study also reported that on-line education blended with face-to-face education still is considered more effective than doing these methods separately. However, learning through Electronic Communities of Practice provides immediate access to information needed on the job and can help to overcome the challenges of varieties in geography, time zone and work setting of public health professionals (Corvey, 2003; Falk and Drayton, 2009; Ho et al., 2010).

Another reason why Electronic Communities of Practice should be considered as a tool to foster the learning process in the professional development of public health professionals is the connection between the human brain and an electronic technology. Olson (2012) conducted a series of four laboratory experiments on the impact of search engines and ready access and retrieval of digitized information on human memory and cognition. They concluded that, "processes of human memory are adapting to the advent of new computing and communication technology" (p. 2). According to Olson, the participants in his study preferred to remember the location of the information rather than the information itself. Electronic Communities of Practice showed potential to enhance learning as it provides the convenience o seeking information through technology (e.g., computers) rather than memorizing it. This conclusion is supported by the previous study by Ho et al. (2010) that Electronic Communities of Practice offer immediate access to information that is needed by professionals and is provided by their peers or by repositories of current and historical discussions. Through

Electronic Communities of Practice, public health professionals select the information they need, which is important for their current professional needs. The rest of the information that relates to their general responsibilities can be stored in their Electronic Communities of Practice and accessed for later use.

Previous studies have discussed specifically how online communities influence professional development especially for health professionals. A study by Corvey (2003) found that nurses who participated in an online community had significant improvement in computer skills, communication skills, and practice-based information. These nurses learned skills through discussion on the listserv, networking with other members of the online community, mentoring others who asked for advice, or just navigating the site. In regard to professional development, Corvey explained that these nurses gained more pride in the profession, became more politically active, more powerful and more critically reflective through their participation in the listserv. Corvey concluded that Electronic Communities of Practice can be a source for continuing professional education of health professionals by promoting both professional practice and professional identity development.

The advancement in technology makes Electronic Communities of Practice more of a standard than the exception for professionals to learn from their peers in solving their daily problems at work. One of the examples of using mobile technology to retrieve information was conducted by Ranson, Boothby, Mazmanian, and Alvanzo (2007). This study was constructed to learn more about the use of personal digital assistance in practice and learning by describing the use of personal digital assistance in patient care and a personal digital assistance version of the Virginia Board of Medicine Continuing Competency and Assessment Form (CCAF). This study demonstrated that the use of personal digital assistance is associated with the value of information for making clinical decisions, in educating patients, and teaching medical students. From the study, it was concluded that the use of personal digital assistance has the potential to foster professional development for health professionals as long as the information is easily accessible and useful for on-the-job practices.

Another study conducted by George (2011) which investigated the evaluation of current technologies in health care revealed that the participants gave uniformly positive evaluations of the mini course. Participants in this study also identified several current tools that were perceived as being potentially useful in their professional lives, including news aggregators, Google Alerts, and social networking sites, such as Facebook. This study suggested that social media technologies will be crucial in helping health professions to adapt to a new, networked era if used responsibly.

Ho et al. (2010) gave another example of the Electronic Communities of Practice in helping to foster professional development in their article. The experiment compared Academic Detailing (AD) that was administered through Technology-Enabled Academic Detailing (TEAD) and face-to-face AD session. The purpose of this study was to determine the effects of both methods on the care of patients with diabetes in urban and rural communities. The study found that knowledge sharing occurred through TEAD sessions helped physicians seek additional and personalized information for pharmacists beyond the limited time of the face-to-face AD session. Participants in TEAD were satisfied, and TEAD was effective in developing interprofessional Electronic Communities of Practice. Through Electronic Communities of Practice, health practitioners were able to maintain their own identities in practice as they helped each other and made decisions collectively about the adoption of evidence to a certain degree into clinical practice based on varying circumstances (Ho et al., 2010).

These trends show how much potential exists for professionals to stay up-to-date with the world around them even while doing their daily activities through their mobile devices. The online sites that were created for online communities through computers now can be delivered to the members of the community on-the-go. Although it is true that the participation will be more difficult to measure with these current trends, learning processes will be enhanced. Professionals can share their thoughts and questions, and they can receive immediate responses from their social network. They can be made aware of events and opportunities that are offered by professional agencies or other educational providers quickly through emails or notification features on their mobile devices. Even conferences or seminars, such as the APHA, have an application that can be installed on mobile devices with just one click. This application allows attendees to choose sessions to attend and functions as their personal schedule for the APHA.

Currently, professionals have many options for finding engagement in the online community, either professionally or for personal use. Online networking sites such as Facebook, Myspace, and Twitter allow individuals to share ideas and information outside the boundaries of a profession. Current technology also allows online networking to be easily and readily accessible through mobile devices, such as tablets, mobile telephone, Personal Digital Assistants (PDAs), and other devices which may lack the convenience of a large screen, mouse, and keyboard. Although participation in Electronic Community of Practice will be more difficult to measure with these current trends, the learning process could be more enhanced. Through Electronic Communities of Practice, professionals could share their thoughts and questions, and they could get immediate responses from their peers through mobile devices. They could be aware of events and opportunities offered by professional agencies or other educational providers quickly through push emails or notification features. Even conferences or seminars, such as the ones organized by the APHA, have an application that can be installed on a mobile device with just one click. This application will allow attendees to choose sessions to attend and act as their personal schedule manager for APHA.

The astonishing growth of assessing information through mobile devices has brought many benefits, but we still need to pay more attention to what information is being acquired, where it comes from, what the information is being used for, and how the information is affecting us as individuals and as a field. Although no guarantees exist that Electronic Communities of Practice could maintain the quality of the information stored in their sites, the professional membership of these communities reflects the quality of information provided by its members. The factors that influence professional development through Electronic Communities of Practice have been discussed in the literature (Falk and Drayton, 2009). The most discussed features of Electronic Communities of Practice that can enhance learning are the level of participation, accessibility of information provided, the design and features offered, and the membership types (Falk and Drayton, 2009).

According to Falk and Drayton, regarding the level of participation, it is important to remember that the retention in participating in an online community depends on the motivation of individuals since there is no credit offered for participating. Participants in online communities are both recipients and providers of professional development. As providers of Electronic Communities of Practice, the members have the same right and freedom in providing, leading, and facilitating the discussion on Electronic Communities of Practice. The way the online community members participate, inhabit, and learn, have implications on the internal and external processes of change and growth. Electronic Communities of Practice enable sharing knowledge, so members can learn from each other about information, techniques, and subject matters in their work or applicable to be applied to their work.

Another factor is the accessibility of information that is provided on the site. Falk and Drayton (2009) used the term "implementation metaphors" to describe how members of Electronic Communities of Practice can access and put the information into a relevant context for their learning needs. For example, for face-to-face conference activities, members expect interaction with peers with similar interests, poster presentations, and keynote speakers.

The next factor is the design and features of the sites. Professional development will be more rapidly enhanced if the site allows the members to choose their own preferences based on their own needs (Falk and Drayton, 2009). They stated that, " the increased capabilities to combine features that optimize content retrieval, content creation, and collaboration, and to customize users' experience according to their preferences, history, and community affiliation, have created new possibilities that must be taken into account when creating learning communities for professional development" (p. 19). The statement warrants the importance of the structure of the site, tools, model of interaction and administrative structure of online community in order to have a significant effect on the professional development of the members.

For the type of membership factor, building trust and mutual knowledge develop more quickly in an online community in which the membership is restricted (Falk and Drayton, 2009). For a heterogeneous membership, the likelihood of brokering expertise, ideas, and tools among participants will increase, and this difference can encourage the learning process among members. The criteria of participation, either restricted or open, shape how the community will form and evolve.

However, many studies also stated that for the members of an online community to realize the potential of professional growth through their participation, some crucial changes are needed (Ho et al., 2010). First, in taking advantage of conversation and exchange of information, a change of consciousness is needed. Second, in the process of building trust, self-presentation, exposure control or protection, constructive exchange development, and cultural exchange are required. Third, the community needs to use explicitly the power to distribute knowledge, share resources and use appropriate tools to the professional practice of its members.

Sargeant et al. (2004) suggested the following guidelines to enable a successful Electronic Community of Practice:

1. Members should be voluntarily involved and self-organized in order to enhance their learning;

2. Electronic Communities of Practice should facilitate the relationships and creativity among members;

3. Electronic Communities of Practice should focus on problems in order to generate solutions from multiple perspectives;

4. Each member has the same right to become a leader, have access to transparency, and have public accountability and freedom to experiment and succeed within the boundaries of Electronic Communities of Practice.

Sargeant et al. (2004) also recommend that members always have access to Electronic Communities of Practice, have shared identity as members of Electronic Communities of Practice that support collaborative problem solving, and maintain the growth and sustainability of these communities.

Both face-to-face Communities of Practice and Electronic Communities of Practice can be formed formally by organizations or professional associations or can be formed informally by individuals within an organization. These are formed for many different reasons. Communities of Practice were formed to train members to become professionals (Cope, Cuthbertson, and Stoddart, 2000; Lindsay, 2000; Plack, 2003). Others were formed to share knowledge between professionals (Pereles et al., 2002; Richardson and Cooper, 2003; Russell, Greenhalgh, Boynton, and Rigby, 2004). Another reason to form Communities of Practice was to complete a certain task (Gabbay et al., 2003; Lathlean and May, 2002; Wild, Richmond, de Mero, and Smith, 2004).

Communities of Practice are considered formal when they are intentionally created by professional organizations, educational providers, or organizations to facilitate their members in serving the Communities of Practice's purpose (i.e., become professionals, share knowledge, or finish a task). Some examples of these formal communities of practice are *ph-Connect* through the Centers for Disease Control and Prevention (CDC) website, *APHA-Connect* through membership with APHA, and *member communities* through SOPHE. Although some of these memberships of Communities of Practice are free, some of them require membership fees. This is another challenge for public health professional to interact with their peers through formal Communities of Practice. In addition to that, most of these mainly serve as an announcement board for the members. Lack of interaction within these communities is one of the main reasons to promote informal learning in the workplace.

The members of the Communities of Practice meet on a regular basis. During these interactions, professionals share information, insight, and advice in solving each other's problems at work. Tools, standards, designs, manuals, and other documents are the results of such discussion. As members interact to discuss common problems, they develop an understanding, implicit or explicit, of the problem that may result in the personal satisfaction of having a sense of belonging to an interesting group of people (Wenger, McDermott and Snyder, 2002). These professionals may even develop a common sense of identity over time through which they construct a Community of Practice. All of these experiences as members of Communities of Practice support the public health professional's development.

3.4 CHALLENGES IN FORMAL AND INFORMAL LEARNING

Despite the advantages offered by CPE, a study by Johnson, Glascoff, Lovelace, Bibeau, and Tyler (2005) found that over 60% of public health professionals did not conduct research or participate in professional development activities due to various reasons. The heavy workload, cost to participate, lack of administrative support, child care and home responsibilities are some of the barriers that resulted in low participation in CPE activities (Bower, Choi, Becker, and Girard, 2007; Schweitzer and Krassa, 2010). Due to the multiple challenges of attending CPE and the advantages of learning in real-time situations in dealing with vast public health issues in the community, the idea of informal and incidental learning is introduced as a significant aspect of continuous learning for public health professionals that complement the formal learning in CPE.

The need for specific competencies in the CPE was stated in several research. Demers and Mamary (2008) found that organizational development, evaluation, management, policy, and advocacy are the primary areas needed to be included in CPE. A study by Price et al. (2004) found that administration, evaluation and applying appropriate research principles are those needed in CPE. Davidson (2008) found that public health professionals need additional training in designing data collection instruments, securing fiscal resources, interpreting evaluation and research results, carrying out evaluation and research plans, and developing plans for evaluation and research. All these needs were emerging because public health professionals observe the gaps between their current skills and the skills needed to perform their jobs.

However, CPE is targeted for a single profession and this has become a challenge in the public health profession. Nurses, physicians, pharmacists, and other professionals in public health have different needs in doing their job; therefore, it is hard to fill these needs in one single CPE. Another issue is the time needed for public health professionals to transfer the skill gained from participating in CPE on their job. This issue is arising because of the complexity of changing the health behavior of the community.

The first problem with formal and mandated CPE activities is the time constraint of public health professionals. A study by Johnson et al. (2005) found that over 60% of public health professionals did not conduct research or participate in professional development activities due to various reasons. The heavy workload of the professionals in the study was shown to be the most significant cause of low participation in CPE activities (Johnson et al., 2005).

In addition to the time constraint of public health professionals, financial constraint is also another reason for the low participation in CPE activities (Demers and Mamary, 2008). Most CPE activities require participants to become members of their organization before they register for the program. The cost of membership and registration may not be covered by their home institution. A study by Demers and Mamary (2008) revealed that financial constraint was the primary reason for public health professionals' lack of participation in CPE activities. Public health professionals stated that their employer gave them the opportunity to attend CPE activities; however, 38% of public health professionals stated that they were not reimbursed for the money they spent to attend CPE activities.

Another problem besides resource constraint for participating in CPE relates to the intention of public health professionals to participate in CPE activities. As mentioned earlier, some of these "formal and mandated" CPE activities offered Continuing Education Contact Hours (CECH) that can be used to maintain professional certification of public health professionals. Therefore, public health professionals may only participate in CPE to acquire or maintain their credentials as professionals (Guskey, 2000). For instance, CHES and MCHES credentials need to submit at least 75 contact hours during a five-year period of re-certification in addition to the heavy workload in their professional positions in order to maintain their professional credentials. Attending these seminars and conferences is the only way public health educators can acquire these 75 contact hours. CPE is mostly considered as a means to get points on contact hours to professional re-certification. As Marsick and Watkins (1990) stated that professionals ready to learn when they are "at the point-of-scale, so to speak, yet training and development is often treated as a commodity for which employees are scheduled at the convenience of the organization" (p. 3). Consequently, there is no guarantee that learning is taking place during public health educators' limited participation in CPE activities.

The next problem with CPE relates to the different needs of public health professionals from CPE activities. Most CPE is targeted for a single profession, and this fact has become a challenge for public health professions. Public health practice is an activity rather than a specific discipline, and public health efforts are conducted by people from diverse backgrounds, ranging from health communications to those whose work in public policy (DiClemente, Salazar, and Crosby, 2013). Public health professionals are comprised of individuals from various backgrounds and at various levels of understanding of certain public health issues. Nurses, physicians, pharmacists, and other professions in public health have different needs related to their jobs; therefore, it is almost impossible to fill these needs in one CPE activity. Previous researchers have shown that different skills are needed by public health professionals in CPE activities. Demers and Mamary (2008) found that organizational development, evaluation, management, policy, and advocacy are the areas that need to be included in a CPE. A study by Price et al. (2004) found that administration, evaluation, and application of appropriate research principles are needed in CPE. Davidson (2008) found that public health professionals need additional training in designing data collection instruments, securing fiscal resources, interpreting evaluation and research results, carrying out evaluation and research plans, and developing plans for evaluation and research. All these needs were emerging because public health professionals observe the gaps between their current skills and skills that are needed to perform their jobs. It is a challenge to include all these needs in one single CPE. Davidson (2008) concludes that "providers assume professionals need to acquire knowledge solely to problem-solve predictable issues at work" (p. 9). This assumption of public health professionals' needs influences the general type of knowledge given in CPE.

The next problem is the actual impact of CPE activities to improve public health professionals' job performance. Although professionals can learn through various methods, limited time in CPE only allows the learning through instruction and may be unable to facilitate fully on each inquiry that public health professionals might have. CPE has limitations in terms of evaluating the impact of learning in the actual performance of participants. CPE is considered to be an inert form of education and training for professionals (Beckett, 2001). Beckett also stated little evidence supports the transition of the learning process from the CPE classroom to the workplace. Furthermore, powerful learning occurs when professionals learn by doing in their own workplace.

In addition, learning by doing in the workplace is an effective strategy for professionals to continue learning in their profession. These workers must attend CPE offerings in their own workforce to ensure that they keep abreast of evolving organizational, ethical, and communication concerns within their Community of Practice (Gebbie et al., 2003b).

These findings were supported by Cervero (2003) who stated that we enter a third era of CPE in which education and learning occur in a place and a time when professionals "are most likely to have a need for better ways to think about what they do" (p. 16). For instance, Cervero (2003) suggested that educational activities should allow real-time interactions of health professionals with their clients in their geographical and professional boundaries. Public health professionals need to deal with a variety of cases within their own community geographically and professionally as defined by their Community of Practice.

Evidence of the benefit and the challenges in attending CPE programs and the evidence of effective learning in the workplace have led many researchers to explore all possibilities in continuous learning in the workplace to benefit public health professionals. These efforts should include the professionals in designing their own learning goals to optimize their professional practice in the workplace. Public health professionals are adult learners, and their learning should progress in agreement with adult education perspectives which are learning from experience, learning informally, and learning from others (Desikan, 2009).

Challenges may also arise in the informal learning, whether in face-to-face settings or in an Electronic Community of Practice. In Communities of Practice, these issues include: distance, size, affiliation, and culture. Wenger, McDermott and Snyder (2002) describe how distance can be a challenge in Electronic Communities of Practice since members mostly interact with other members through Web conference, telephone, or other means. There are connections and visibility issues between members in their interaction. Another factor according to Wenger and coworkers is size in terms of knowing the other members of an Electronic Community of Practice personally. This challenge emerges especially in Electronic Communities of Practice which have a large number of memberships. The next issue, according to Wenger and coworkers, is affiliation as the distribution of knowledge inside an Electronic community of Practice should not have the adverse effect on a member's company. The difference in cultural backgrounds can be another challenge when members have difficulties relating to one another when communicating through Electronic Communities of Practice. This difficulty to relate to others can lead to communication difficulties and to misinterpretation among members.

Wenger, McDermott and Snyder (2002) acknowledge, however, that there are some downsides of Communities of Practice. Successful communities should acknowledge and leverage this awareness in order to support the growth and ensure vitality over the long term. These downsides may come from individuals as members, the multiple communities, and the barrier from members' organizations. The simplest downside is that the community may not be functioning well due to the temptation of ownership/imperialism of domain, or the community creates cliques outside the focus of Communities of Practice, or when a community loses its competence in supporting the professional development of the members.

These downsides can be overcome by addressing each of the fundamental elements of the Communities of Practice. In the domain elements, these downsides can be overcome by: (1) establishing the legitimacy and strategic value of the domain; (2) clarifying the link to business matters and finding ways for the community to add value; (3) offering inspiring challenges; (4) including the community in important decisions; (5) holding the community accountable for the reputation of the firm in the domain; or (6) exposing the domain to other perspectives (Wenger, McDermott and Snyder, 2002).

In terms of the community elements, Wenger and coworkers proposed the following to overcome the downsides: (1) engaging the community in shared problem solving; (2) involving new generations in the community, and (3) connecting the community with other communities. In terms of practice elements, they recommended the following as treatments for the downsides: (1) encouraging member involvement in the development of the practice by making enough time to participate actively; (2) balancing joint activities with the production of artifacts; (3) initiating exciting knowledge-development projects; (4) benchmarking the practice of other communities that include competitors; (5) challenging members to help teams with leading-edge issues; and (6) valuing member participation by allowing their contributions to build their reputation and affect their positions in the organization.

Living with downsides is not easy, but Electronic Communities of Practice also offer the advantages of supporting the members' professional development. According to Falk and Drayton (2009, p. 20), the following decisions will shape the nature of the participants' experiences in their professional development:

1. Will the users anticipate the professional experience to be more similar to visiting a specialized library or to attending a meeting with a colleague?

2. What role will a specific corpus of content play in the participants' professional experiences?

3. Will the site result in a growing knowledge base, resource center, or library that users can read, share, and contribute to?

4. Is the site intended to provide rich virtual meeting space where ideas are exchanged and collaborative work is done, but where no specific product is necessarily anticipated?

Therefore, Wenger and coworkers also provided ways in order to cultivate Electronic Communities of Practice to minimize the challenges that members faced. According to Wenger, McDermott and Snyder (2002), several steps need to be taken to define a distributed community:

1. Achieve stakeholder alignment.

2. Create a structure that promotes both local variations and global connections by combining diversity and connection, connecting people, and avoiding hierarchy.

3. Build a rhythm to maintain community visibility by arranging teleconferences, organize face-to-face meetings, facilitate threaded discussions, link modes of interaction, and make judicious use of broadcast technology.

4. Develop the private space of the community through personalized membership, small group projects and meetings, having organized or impromptu site visits, and remain opportunistic about chances to interact.

Regarding the type of interaction, a study conducted by Ranmuthugala et al. (2011) found that 11 out of 32 studies revealed that health professionals commonly used face-to-face interaction in the workplace and only a few of them use technology as medium. The study suggested that geographical distribution of members and objective of Communities of Practice influenced the type of interaction between members. These interactions in Communities of Practice that resulted in the informal learning of professionals are encouraged to create changes in job performance of public health professionals.

CHAPTER 4

Learning Organization for Professional Development

4.1 ORGANIZATION: GROUP AND TEAM

An organization is a place for a group of people who work coordinately to achieve a common goal. In each organization there is a team or group which is formed to complete a particular task. A collection of two or more people who work with each other regularly and have common goals, shared expectations, joint activities, and mutually agreed norms are generally defined as a group (Schermerhorn et al., 2002). Groups can be defined as people who have similarities, complement each other, commit to achieve goals determined by the leader and are willing to be responsible towards the leader. The strength of the group lies in the commitment of its members to follow command or direction from the leader (Robbins, 2003). Groups are often permanent but are possible to change when the organization rearranges its structure. On the other hand, a *team* is very different from a *group*. A team is a small group of people with complementary skills who are committed to shared intentions, who obtain goals, and who share a common approach where they commit themselves to share responsibility (Robbins, 2003). Goals will not be achieved if one of the team members is unable to attend a meeting because one member cannot replace another member. So the strength of a team lies in the similarity of goals and interpersonal relationships.

A team can be classified according to its purpose. There are four general forms of the team which can be found in everyday life. According to Robbins (2003), team types are problem-solving, self-managed work, cross-functional, and virtual teams.

Problem-solving teams are formed specifically to solve a problem in the organization. Self-managed work teams are interconnected to determine, plan, and manage their activities and duties under reduced or no supervision (Figure 4.1). Cross-functional teams may have different duties or functional expertise working together toward a common goal in the organization. Virtual teams work together from different geographic locations and usually rely on communication technology such as email, fax, and video or voice conferencing services (Figure 4.2).

Each team or group in the organization has group norms. Group norms are unwritten rules which define group role behavior that members can accept. Norms include the level of performance assessed by the group, the performance of members in the group, the relationships with managers, and other formal aspects of organizations. Group norms are obtained and learned from social processes (Champoux, 2010).

FIGURE 4.1   Self-managed work team.

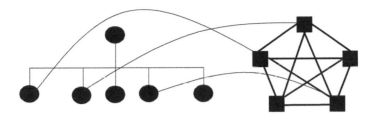

FIGURE 4.2   Cross-functional work team.

Adapting group members to these norms is part of the price that must be paid as a result of being accepted as a group member (Jewell and Siegall, 1990). Conformity to the group norms that are possible is compliance. Compliance means that someone goes according to the norm in the group, but does not accept the norm. The person may be obedient to help groups appear united to outsiders or to prevent conflict in groups. Conformity to group norms can lead to more effective group performance (Champoux, 2010).

Groups in an organization go through several stages of development, namely forming, storming, norming, performing, and adjourning.

4.1.1   Forming (Group Formation Stage)

The first stage is the forming stage, characterized by many uncertainties related to goals, structure, and leadership in groups (Champoux, 2010). Group members meet each other for the first time and introduce themselves to each other. At this stage, group members begin to identify a number of questions about what the group can offer them, what kind of contributions they can give to the group, and if it is possible that their needs will be fulfilled at the same time they contribute to the group. Group members identify the possible rewards they receive if they give maximum contribution to the main tasks and functions that have been given, whether they are in accordance with the expectations of each group member. At this stage, the members are interested in getting to know each other and finding what is considered acceptable behavior in defining rules in groups (Schermerhorn et al., 2002).

This forming phase will be resolved after each member begins to know his position as a group member.

4.1.2 Storming (Intragroup Conflict Stage)

Storming is the stage of group development in periods of high emotionality and tension among group members. In this stage, hostility and conflict can occur and there are usually many changes. Each group member seeks to impose his will to achieve status and a strategic position in the group (Schermerhorn et al., 2010). Periods of conflict can interfere with interpersonal relationships. Group members show each other their existence in carrying out tasks to get status. The implementation of the task seems to be running alone and on its own without any clear direction because there is no mutual cooperation between group members. This stage will be resolved if there is a relatively clear hierarchy of leadership in groups, and group members are oriented towards problem solving (Champoux, 2010).

4.1.3 Norming (Group Cohesion Stage)

Norming is the stage of group cohesion. This stage expects that group cohesion has been formed. The norms received must be defined and agreed. To achieve the goals of the group, it is necessary to establish norms and procedures that have been agreed upon together. Group members have been able to define their role and its relationship in the group so that they can freely share ideas and solutions for improvements. There is already a clear direction in carrying out the tasks of each role in the group.

During the stage of group cohesion (norming), the group agrees about the right behaviors of the members. Members accept each other, and an identified culture of a group emerges. Conflict is less intense than in the previous stage. When the conflict is accepted as part of group norms, the group defines acceptable conflict behavior. Conflict at this stage focuses less on the social structure of the group than on different ways of doing group assignments. The way a group member works on a task is evaluated. Conflicts can arise if someone sharply deviates from group norms about task behavior (Champoux, 2010).

4.1.4 Performing (Task Orientation Stage)

During the performing or task orientation stage, the group has reached organizational maturity and is functioning properly. Group members feel comfortable with each other in group task orientation. Group members set common goals and implement division of work in groups. Tasks that have been divided according to the roles in public health services are carried out and each group member focuses to complete group work (Champoux, 2010). This stage transforms group members from knowing and understanding each other to a single point of synergy to complete group tasks (Robbins and Judge, 2013).

Group members enjoy each other's group assignments and accept group norms. Group members agree on common goals and regulate the division of work in the group. Tasks that must be completed are carried out and each group member focuses to complete group work (Champoux, 2010). At this point, the group is expected to have organizational maturity and to be functioning properly. The existence of conflicts in the group can be resolved in a creative way, the group has a constant structure, and group members are motivated by common goals. The main challenge is ongoing efforts to improve relationships and performance. Group members must be able to maintain interpersonal relationships in addition to providing the best performance. Group members must be able to adapt because opportunities and demands change over time (Schermerhorn et al., 2002).

4.1.5 Adjourning (Termination Stage)

Adjourning or termination stage is the stage of group development after reaching the goal, then dissolving and ending existence as a group that can be identified (the stage of termination or suspension). The group then redefines its duties and membership so that the group returns to the first stage of development and repeats the process.

Functional groups and informal cohesive groups achieve task orientation at the same stage. Under certain conditions, the group repeats the stages and experiences (Champoux, 2010) (Figure 4.3).

The effectiveness of a team's performance can be assessed from the context, composition, and processes that occur (Robbins, 2003). Four contextual factors that are significantly related to team performance are adequate resources, effective leadership, climate of trust, and performance evaluation and reward systems that reflect the contribution of a team (Robbins, 2003).

   a. Adequate Resources: The performance of a team is very dependent on the resources they have. Inadequate resource conditions can reduce the team's ability to carry out work effectively including achieving goals. One important factor for the effectiveness of group performance is organizational support for the team including timely information support, adequate equipment, adequate team members (employer/ employees), encouragement, and administrative assistance.

   b. Leadership and Structure: Leadership in a team plays an important role in empowering the team by delegating responsibilities to team members and as facilitators, by ensuring that the team works well together and reducing interpersonal conflict. If a team has a leader and a clear structure, a clear division of work will be realized.

   c. Climate of Trust: The climate of trust in the team will facilitate and give birth to good cooperation, and reduce

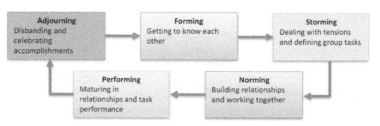

FIGURE 4.3   Stages of group development.

suspicion while ensuring that teammates will not take advantage of it. Team members who have high mutual trust will be able to work collectively so that they tend to get good outcomes in completing their work assignments in the organization. In addition, with the trust between members, good communication can be established within the team to help each other in completing tasks.

d. Performance Evaluation and Reward System: Performance evaluation and regular rewards for team members can increase team effectiveness. The team can assess its performance with the feedback provided by its supervisor. This feedback allows the team to identify activities that can be improved or changed to achieve common goals.

Another factor that also affects team effectiveness is composition. Robbins (2003) states that the team composition category includes all matters relating to how should the team place the members, the capabilities and personality of team members, role allocation among team members, diversity of team members, team size, and members' desire for teamwork roles. The descriptions are as follows:

a. Capability of Team Members: Team performance is greatly influenced by the level of knowledge, skills, and abilities of each team member. A team with a high ability to play will provide smart solutions to every complicated problem faced by the team. Smart leaders can help team members who have less ability to carry out their duties. In this case, if the abilities of team members are uneven, the leader's task is to make a decision to pair the team members who have less ability with other team members who have more ability in each delegation of tasks.

b. Personality of Team Members: Personality can influence the work behavior of individuals in the organization. A team that has members with good personality will respect each other and have good acceptance to their teammates so that an effective work climate will be created. On the other hand, if there is a team member who does not have a good personality, it will give a negative impact on teamwork and team performance because respect is one of the human basic social needs.

c. Allocation of Roles among Team Members: Allocation of roles and the division of workload or tasks fairly is important in order to optimize team performance and avoid interpersonal conflict.

d. Diversity of Team Members: Diversity is a condition where team members from one work unit have different demographic characteristics including age, gender, race, education level, etc. Diversity in teams must be managed properly so that it becomes a valuable asset for the organization, especially in terms of problem solving. An organization consisting of members with different experiences has the potential to add strength to the team, especially in increasing the point of view on problem solving and brainstorming.

e. Size of Team: Several studies have shown that maintaining a team with a small membership is the key to increasing team effectiveness in coordination. Conversely, a team that has a large number of members is considered to reduce cohesiveness and communication, and increase laziness of members in performing their tasks. In this case, team members will be difficult to coordinate and poorly communicate with each other. As a result, team performances will be ineffective.

f. Member Choice: Another important factor related to the level of team effectiveness is the selection of team members who can commit to the team and members who can complete the required tasks.

In order to create an effective team in the organization, it also requires a directed process. This process includes activities

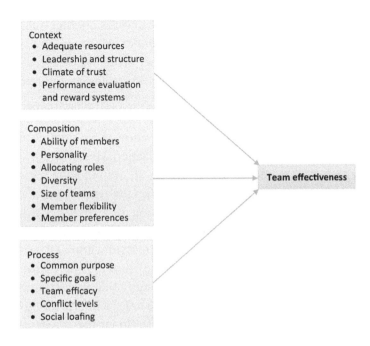

Context
• Adequate resources
• Leadership and structure
• Climate of trust
• Performance evaluation and reward systems

Composition
• Ability of members
• Personality
• Allocating roles
• Diversity
• Size of teams
• Member flexibility
• Member preferences

Team effectiveness

Process
• Common purpose
• Specific goals
• Team efficacy
• Conflict levels
• Social loafing

FIGURE 4.4   Team effectiveness model. (From Robbins, S. P. and Judge, T. A. 2013. *Organizational Behavior Edition 15*. New Jersey: Pearson Education.)

to achieve common purpose and specific goals. The process in handling conflicts and social loafers will also affect the effectiveness of a team (Figure 4.4).

4.2   ORGANIZATIONAL BEHAVIOR AND CREATING EFFECTIVE ORGANIZATIONS

Group dynamics is a strength that operates within groups. It affects the way members communicate and work with one another. The ego that lies within group members is one of many factors that is underlying the emergence of group dynamics. The existence of ego within respective members cannot be opposite with norms and rules that have been applied within an organization. The objectives of group dynamics are meant to raise self-awareness within a group member, thus the sense of respect towards another member can be grown. Furthermore, group dynamics helps to grow a sense of solidarity among members so that they are able to respect each opinion, fosters open communication, and creates good intention among group members.

Group dynamics is necessary since member behavior within a group will highly determine the effectiveness of an organization. According to a number of experts, organizational behavior includes some matters that affect how persons, as an individual or a group, behave within an organization, and its effect towards organizational structure and system as well. Diversion of people's attitudes and behaviors within an organization is studied to find out the solution of how management manages an organization effectively. Conceptually, Robbins and Judge (2013) provide the definition of organizational behavior as a field of study that investigates the effect of individual, group, and structure toward organization and then implements it for the sake of organizational effectiveness.

Organizational behavior studies three determinants within organization: individual, group, and structure. It implements knowledge of behavior that is linked with activity and result of work from organization members. There are two areas of focus in organizational behavior: actions and attitudes of organization members (Ratmawati dan Herachwati, 2007). George and Jones (2002) stated that organizational behavior is required as a study about various factors affecting individual and group actions within an organization, and also how organization maintain its environment. Similar to George and Jones, Robbins and Judge (2013) and Davis (2002) provide a depiction that this organizational behavior study acquires a set of tool (concepts and theories) that helps people comprehend, analyze, and explain behaviors within organization.

For a manager, studying organizational behavior helps refine, encourage, or change work behavior, in either individual, group, or organization holistically hence the organization is able to achieve setting goals.

Accordingly, it can be said that organizational behavior is highly focused on the "Human Side of Management," thus this approach in a management field is a behavioral approach (behavioral approach to management). The knowledge that is obtained by learning this behavioral organization will help managers in identifying problems, determining the correct way to solve them, and discovering that changes will make a difference, by using a behavioral approach towards an effective and efficient organizational performance (Supartha and Urge Ketut Sintaasih, 2017).

Organization is defined as resources configuration within a work activity, where each activity has been arranged systematically to achieve the determined objectives. An organizational objective can be achieved immediately if the organization is able to create an effective one. According to Sedarmayanti (2009), organizational effectivity is a parameter of organizational success in achieving goal/objectives.

During the journey of making an effective organization, there will always be obstacles that get in the way towards organization performance. A high performance organization will always strive to overcome the obstacles through commitment made by the organization's top managers.

The parameter of organizational success lies in how the organization manages to control or maintain any challenges that it faces. Some challenges are as follows:

1. An effective or high-performance organization always employs an open system that is influenced by external social changes, which focus on the global condition and customer expectation that is always changing in a rapid period.

2. The hardest challenge in actualizing an effective organization is the way of integrating the five components of a High Performance Organization (HPO) in its activity and function.

3. The role of middle managers is also a challenge in creating an effective organization. They are often required to help apply one

or more component of HPO in order to run their organization along with their process to become an effective organization.

4. The first challenge for top managers is their decision and commitment in implementing HPO.

5. The last challenge is to start an organization with high performance whether it be a new organization starting from zero (Greenfield) or redesigning the existing organization. In his book, Schermerhorn (2002) stated that the organization that applies a new design has experienced a financial increase of approximately 10% per year. However, the organization that has undertaken redesign has experienced a profit increase of 6.8% per year. Moreover, the organization that has not undertaken any refinement has experienced a profit improvement of 3.8% per year. It can be concluded that even though the three design types have experienced financial increase in responding to external and internal pressure, the HPO design has shown the best result. Furthermore, starting HPO from zero will lead to a maximum result compared to undertaking changes in an existing organization.

Any means that is required to create an effective organization should consider the following elements:

1. Strategy: Strategy means a role, objective, and a strategic direction that summarizes work result of organization and/ or division, thus it becomes clear and appropriate.

2. Structure, Capacity and Capability: This element refers to the people who are able to undertake the right job through "fit to objectives" structure and clear depiction of accountability and role relation.

3. Leadership: This element refers to a leader who possesses the ability and capacity to encourage sustainable successful business.

4. People Systems and Processes: A leader needs to be supported by a good system and process. The system and process work within the organization to send a message, share information, and make a good decision in every business. Organizational process and system are the extension of leadership, creating consistency and trust.

5. Employee Engagement: Employee engagement explains the rate of the employees' involvement, involving heart and mind that are harmonious with their work and the organization they work at. The employees who are involved have the following characteristics:

   a. They are satisfied with their current job and the organization as their employer.

   b. They are committed to making their job and organization succeed.

   c. They are proud of their organization and their job.

   d. They are willing to talk about their job and organization positively.

6. Customer Experience: High customer satisfaction level and loyalty can be obtained through the employees who are aware about customer needs, who act based on customer feedback, and who are supported to meet customer needs.

The six organizational key elements are necessary to encourage employee involvement in an organization. This involvement affects customer experience and finally leads to overall organizational performance in terms of high productivity and profitability (Haid et al., 2010).

The integration among the first five elements has a strong and beneficial result: a culture that is based on customer-focused performance. There is no single initiative that is able to emerge the organization's effectiveness. Excellence is required in every effective organizational frame element to fit in a competition (Figure 4.5).

122 ■ Professional Development and CPE

FIGURE 4.5 Right management's organizational effectiveness framework.

The effective organization is scored based on a number of criteria, either specific or general, such as input process and output process. The coordination effort is undertaken by a manager including planning, organizing, leading, and controlling the organization member's behavior. The main factor that determines the individual and group relation is task and authority. Therefore, a manager is required to design an organization's structure and process in order to facilitate communication among organization members (Schermerhorn, 2002).

Approaches for organizational effectiveness include:

1. Goal/Objective Approach: According to this approach, an organization exists in order to achieve goals. Effectivity is a goal achieved with cooperative effort. This goal-achieving

rate indicates the effectivity level. The higher the achievement of an organization's goals, the higher its effectivity level. One of the practices that are commonly applied is objective-based management.

2. System Theory Approach: According to the system theory approach, a system categorizes elements that are individually correlated with one another and that interact within the environment as an individual or collective. A manager within the organization employs a system concept to monitor the internal and external world, and how the parts interact with and correlate to each other as well.

3. Stakeholder Approach: A stakeholder approach's objective is to achieve balance between many system parts and organizational constituent interest satisfaction, all individuals and individual groups that have interest within the organization. These consist of employees, customers, shareholders, directors, suppliers, creditors, and officials on all government levels, organizational managers and the public. Individuals and groups expect that the organization run in a way that fits along with their benefits.

The important patterns within an organizational change consist of:

1. The successful change is linked with a multi-step process that creates power and motivation to continue.

2. The change process is encouraged by a top-quality leader who contributes eternal influence towards current changes. This leader forms direction linear with his vision and inspires people to overcome political, personal, and bureaucratic barriers.

4.3 LEARNING CULTURE AND LEARNING ORGANIZATION

A culture is passed on and shared between members of an organization and is manifested mostly unconsciously in every aspect of an organization's life, from the rituals of celebration to how decisions are made (Gill, 2010). A culture in an organization is shown from the way individuals communicate with each other, the type of leadership, the performance evaluation conducted, the physical environment of the workplace, and the knowledge management in an organization (Gill, 2010). According to Schein, 1985 (as cited in Gill, 2010), organizational culture is "the values, basic assumptions, beliefs, expected behaviors, and norms, of an organization; the aspects of an organization that affect how people think, feel, and act" (p. 5). This organizational culture has an impact on creating and sustaining learning in an organization over time.

A learning culture is a learning that is manifested in every aspect of organizational life (Gill, 2010). According to Gill, "learning culture occurs in an organization that continually makes reflection, feedback, and sharing of knowledge as part of its function in a daily basis" (pp. 5, 49). Gill said that a culture of learning is "an environment that supports and encourages the collective discovery, sharing and application of knowledge…. [in which individuals are] continuously developing new knowledge together and applying collective knowledge to problems and needs" (p. 5). The culture of learning contributes to the improvement of capacity of an organization. Gill added that "an organization with a learning culture encourages surfacing, noticing, gathering, sharing, and applying new knowledge" (p. 29). According to Gill, a learning culture in an organization can be developed through continuous individual, team, organizational, and community feedback and reflection. Gill said, "learning should be manifested in every aspect of organizational life where members are continuously learning as individuals, in teams or groups, as a whole organization, in relation to their communities" (p. 49). Watkins and Marsick (1993) stated that learning is often shared informally between individuals that may belong to many different groups or teams. These individuals learn in the workplace as they work together to achieve certain goals through interaction with their peers as they help each other to solve their work-related problems. According to Watkins and Marsick, these collaboration may transform the knowledge of these professionals involved and lead to a learning process at the organizational level that is more difficult to manage or predict. However, a learning process at the organizational level is necessary to facilitate changes in the job performances of these professionals. Thus, they concluded that although learning at the individual level is necessary, it will not be sufficient to influence changes in performance without the ongoing support from systems, practices, and structures in the workplace.

There are, however, barriers to a learning culture that can be manifested in subtle and not-so-subtle resistance. Gill (2010, pp. 15–23) listed the following as the barriers to a learning culture in an organization:

1. Program focus where the attention of a member is usually on program delivery and not on the organization's improvement;

2. Limited resources that make members unaware of many learning opportunities that do not require large expenditures of time and money;

3. Work-learning dichotomy when there is an assumption that work and learning are different activities and learning needs to be conducted in a classroom;

4. Passive leadership where staff only will report success because that is what they think the leaders want to hear and do not ask tough questions about the organization;

5. Non-learning culture that closes off communication as well as stifles honest feedback and reflection, and discourages risk taking that can provide the opportunity for learning;

6. Resistance to change as the tendency to maintain the familiar and not take the risk of trying something new and different;

7. Not discussing the undiscussable that prevents information from surfacing in organizations that could be very useful for learning and change;

8. Need for control that prevents members from communicating vital information to another staff member who is not within a particular line of authority;

9. Focus on short-term simple solutions by taking the easy way out and not investing time, effort, resources, and emotion in the big picture and long view;

10. Skilled incompetence where individuals have the natural tendency to avoid embarrassing or threatening interactions with others, or not accepting responsibility for problem situations; and

11. Blame (not gain language) that puts the other person on the defensive and stifles any interest that person might have had in receiving constructive feedback, reflecting on its meaning, and using what he or she has learned to improve the organization.

Gill (2010, p. 47) also described activities that can create and maintain a culture that is conducive to learning in an organization:

1. Make highly visible, dramatic changes that are symbolic, as well as substantive, of a learning culture in the organization;

2. Ensure that values demonstrated in everyday actions are consistent with espoused values of learning and talk about this alignment of values with employees;

3. Assess and compare the perceived current culture with the desired learning culture;

4. Develop a shared plan with board members and staff for what the organization must do to move from the current culture to the desired learning culture;

5. Allow employees to dedicate time to formal and informal learning that will enhance their capacity to do their work effectively;

6. Develop learning events that are explicitly linked to the strategic goals of the organization;

7. Create ceremonies that give recognition to individual and team learning;

8. Make the artifacts of learning visible to employees, such as a library, spaces for formal and informal conversations among employees, benefits that support education, and computer access to just-in-time information;

9. Praise individuals and groups that use learning as one of their indicator of success.

The organization that continually transforms itself through the learning of its members is called a learning organization. Watkins and Marsick (1993) defined a learning organization as "one that learns continuously and transforms itself" (p. 8). This definition is supported by Marquardt (1996), which defined learning organizations as "companies that are continually transforming themselves to better manage knowledge, utilize technology, empower people, and expand learning to better adapt and succeed in the changing environment" (p. 2). In addition, Marquardt said that a learning organization should provide structure for individuals to apply their knowledge while continuing to empower people within and outside the organization for the success of the organization.

Learning organizations are sometimes intertwined with organizational learning. Both are similar but also different in the usage of the term. Denton (1998) described the difference between these two terms. According to Denton (1998), a learning organization is "an organization that practices organizational learning" and organizational learning is "the distinctive organizational behavior that is practiced in a learning organization" (p. 3). According to Dixon (1999), organizational learning results from intentional and planned efforts to learn and it may occur accidentally. However, organizations cannot afford to rely on learning through chance. Organizational learning takes place through learning and interaction between individuals in the organization. According to Probst (1997), "the individual processes and outcomes in the organization are prerequisites for organizational learning and form an important basis for it" (p. 17). Therefore, organizational learning is always unique to an institution with its own capabilities and characteristics.

Although learning organizations and organizational learning are synonyms, a learning organization is an entity, and the organizational learning is a process, or a set of actions (Denton, 1998). Therefore, a learning organization is something that the organization is, and organizational learning is something that the organization does (Denton, 1998). This differentiation aligns with the Marquardt (1996) description that learning organizations focus on what is learning in organization, and organizational learning refers to how organizational learning occurs in the organization. This means that organizational learning is one of the dimensions or elements of a learning organization (Marquardt, 1996).

In order to become a learning organization, Marquardt (1996, pp. 211–215) stated that organizations need to encourage, expect and enhance learning at all levels and listed the following as a key to successful transformation into a learning organization: (1) Establish a strong sense of urgency about becoming a learning organization; (2) Form a powerful coalition pushing for a learning organization; (3) Create the vision of a learning organization; (4) Communicate and practice the vision; (5) Remove obstacles that prevent others from acting on the new vision of a learning organization; (6) Create short-term wins; (7) Consolidate progress achieved and push for continued movement; and (8) Anchor changes in the corporation's culture.

Marsick and Watkins (2003) considered workplace learning as "the little R&D" (p. 133), because most of the learning in organizations is evolving from the work itself where members learn spontaneously and organically. According to Marsick and Watkins (2003), the workplace provides ongoing experimentation where professionals use their everyday experience as learning outcomes that result in changes of knowledge performance. In learning organizations, individuals learns and share their learning experiences with their peers, which serves as a vehicle for learning in groups and the whole organization.

Learning and working are different concepts but are always intertwined because learning is part of work, and work involves learning (Dixon, 1999). Although learning is commonly viewed as an individual activity, learning in the workplace takes place within a social context to promote collaboration between individuals or teams in an organization (Smith and Sadler-Smith, 2006). The learning that takes place among public health professionals should follow Knowle's theory of andragogy that has the following assumptions (Smith and Sadler-Smith, 2006, p. 89):

1. Adults need to know why they need to learn something before undertaking it;

2. Adults' self-concept is one of being responsible for their own decisions;

3. Adults come to a learning experience with greater volume and variety of life experiences than do younger learners;

4. Adults become ready to learn those things that they need to know to cope with their real-life situations;

5. Adults' orientation to learning is life-centered with the potential of some form of payoff in work or personal life; and

6. The most potent motivators for the adult learner are internal pressures such as job satisfaction, self-esteem and quality of life.

In regards to learning in a workplace, Watkins and Marsick (1993) explained that an organization has the role to empower people, integrate quality and quality of work life, and create free space for learning where teams collaborate and share the gains; and where individuals promote inquiry and create continuous learning opportunities in the organization. The knowledge resulted from these learning processes at all levels serves as nutrient that enables the organization to grow as a learning organization (Marquardt, 1996). According to Smith and Sadler-Smith (2006), these three domains of learning are integrated when "the learner achieves a critically reflective state in which he or she is sensitive to why things are being done in a particular way, and is critically reflective before accepting 'given' solutions to problems or methods of practice" (p. 36). Smith and Sadler-Smith explained that there are three forms of workplace learning that are conceptualized by Mezzirow (1991) as the following (Smith and Sadler-Smith, 2006, p. 36):

1. Instrumental learning focuses on learning aimed at skill development and improving productivity;

2. Dialogic learning involves learning about the individuals' organization and their place in it; and

3. Self-reflective learning involves a transformation of the way a person looks at the self and relationships.

4.4   THE DIMENSIONS OF THE LEARNING ORGANIZATION

In an organization, there are three levels of learning: individual learning, group/team learning, and organizational learning. According to Marquardt (1996), "individual learning refers to the change of skills, insights, knowledge, attitudes, and values acquired by a person through self-study, technology-based instruction, insights, and observation; group/team learning refers to the increase in knowledge, skills, and competency which is accomplished by and within groups; and organizational learning represents the enhanced intellectual and productive capability gained through corporate wide commitment and opportunity for continuous improvement" (p. 21-22).

He added that "individual learning is needed for organizational learning since individuals form the units of group and organizations" (p. 32). However, organizational learning differs from the other levels of learning (Marquardt, 1996, p. 22) because:

1. Organizational learning occurs through the shared insights, knowledge and mental models of members of the organization; and

2. Organizational learning builds on past knowledge and experience—that is, on organizational memory which depends on institutional mechanisms (e.g., policies, strategies, and explicit models) used to retain knowledge. This continuous learning in organizations is strategically used to foster the development of organizations through the changes in knowledge, beliefs and behaviors of members.

In order to measure important shifts that influence an individual's learning, Marsick and Watkins developed an instrument called the Dimensions of the Learning Organization Questionnaire (DLOQ) in 1990 based on the model of dimension of a learning organization by Watkins and Marsick (1993) and was built on the idea that changes must occur at every level of learning in the organization (Marsick and Watkins, 2003). These changes then become new practice or routine that can be used by the members of the organization to improve their job performance (Marsick and Watkins, 2003) (Figure 4.6).

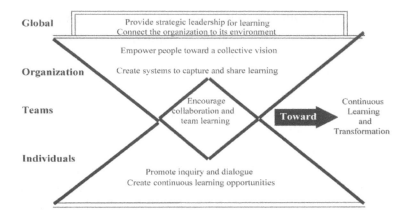

FIGURE 4.6 Model of dimensions of a learning organization. (From Watkins, K. E. and Marsick, V. J. 1993. *Sculpting the Learning Organization: Lessons in the Art and Science of Systemic Change* (1st ed.). San Francisco: Jossey-Bass.)

DLOQ used a six-point scale to distribute responses from "almost never" to "almost always" (Watkins and O'Neil, 2013). To avoid a clustering of the responses, we have used a point scale to measure their responses. Further, DLOQ is a six-point scale therefore respondents can not give two answers for one question, they have select one point for each answer. According to Watkins and O'Neil (2013), items in DLOQ were "vetted with expert and student panels to ensure the language was simple, straightforward, and at an appropriate reading level for a largely professional audience" (p. 137). These items were developed based on seven dimensions of the learning organization: (1) Create continuous learning opportunities (CL), (2) promote inquiry and dialogue (DI), (3) encourage collaboration and team learning (TL), (4) establish systems to capture and share learning (ES), (5) empower people toward a collective vision (EP), (6) connect the organization to its environment (SC), and (7) provide strategic leadership for learning (PL) (Marsick, 2013, p. 130).

DLOQ has been used in many fields to measure organizational capacity to learn and change to increase its overall performance.

DLOQ also has been used in the public health field for the same reason to meet current public health demands. A study by Watkins, Milton, and Kurz (2009) used DLOQ that has been specifically modified to the public health field; it has 65 items which take 10–15 minutes to complete. It consists of the following:

1. Part I (Dimensions of the Learning Organization) in which the participants were asked to think about how their organization supports and uses learning at an individual, team, and organizational level;

2. Part II (Change in Organizational Performance) in which the participants were asked to rate the changes in their organization that occurred in the past year;

3. Part III (Organization Profile) in which the participants were asked to provide the information about their role in the organization and the length of time in their current job position.

DLOQ that was specifically designed for the public health field was used in the study by Watkins, Milton and Kurz (2009) to identify the organizations' capacity to learn and to change to meet current public health demands. This study distributed the DLOQ for the public health field via a commercial survey website to four local public health departments. The findings suggested the DLOQ was also a valid instrument for public sector organizations, and that the learning organization was more correlated with performance than were individual and team learning dimensions.

Like any instrument, DLOQ has its own limitations. According to Marsick and Watkins (2003, p. 138), the following are limitations of DLOQ:

1. DLOQ is a self-reported data and a perceptual measure;

2. DLOQ performance questions often are only answered by middle- and higher-level managers;

3. DLOQ is at best proxy measures for actual performance, and cannot show high and low over time.

However, recent meta-analysis showed that the DLOQ has continued to achieve high reliability for all seven dimensions and show future potential from its increasing use across all of the cited contexts and variations (Song, Chermack, and Kim, 2013; Watkins and Dirani, 2013). Since 2002, there have been 173 requests to use the DLOQ in study in 38 countries, primarily in the United States (63 requests), Europe (35 requests), Africa and the Middle East (27 requests), and Asia (24 requests) (Marsick, 2013). Studies also show that all DLOQ dimensions are able to measure the learning culture in different cultures (Dirani, 2013; Kim and Marsick, 2013). Today, over 70 articles using the DLOQ in many contexts and cultures have been published and have been translated from English to at least 14 other languages (Watkins and O'Neil, 2013). These cumulative works provide evidence that DLOQ demonstrated the validity and reliability in many different contexts and cultures.

Although many learning opportunities are present for professionals to continue learning in the profession, the amount of learning depends on the individuals. Marsick and Watkins (1990) explained that framing and capacity act as delimiters of informal and incidental learning. They added that framing is how individuals relate the selected problem to its context, as individuals explore for interpretations. Capacity is the ability to use the learning over a long period of time and over other learning outcomes. They also described how informal and incidental learning can be enhanced to foster development of professionals.

According to Marsick and Watkins, creativity, proactivity, and critical reflectivity are the enhancers for informal and incidental learning. Marsick and Watkins (1990, pp. 28–31) explained the enhancers as the following: (1) proactivity refers to a readiness to take initiative; (2) critical reflectivity requires people to check out their assumptions before blindly acting on them, pay attention to surprising results and inquire into their meaning, ask probing questions, and reframe their understanding of the problem; (3) creativity enables people to think beyond the point of view they normally hold, helps learners break out of preconceived patterns, and allows people to "play" with ideas so that they can explore possibilities.

4.5 THE IMPACT OF LEARNING ORGANIZATIONS FOR PROFESSIONAL DEVELOPMENT

In a learning organization, individual learning and organizational learning influence each other. Individual learning enhances the organizational learning by scanning the environment and using the information gathered to make a better decision (Watkins and Marsick, 1993). Alternatively, learning organizations encourage public health professionals to put their learning into practice. As Watkins and Marsick (1993) stated, "Learning organizations depend on the participation of many individuals in a collective vision and on the release of the potential locked within them" (p. 195). In addition, Wanto and Suryasaputra (2012) mentioned in their article that a learning culture can be developed in a learning organization to support the continuous learning in the workplace. Literature showed that professionals need to continue to learn in their profession, and organizations need the learning culture to foster development of professionals. However, there are only few studies that have been conducted to understand the role of the learning culture and the participation in professional development of public health professionals.

According to Marquardt (1996), in a learning organization, the transfer of knowledge is indispensable where "knowledge should be disseminated and diffused appropriately and quickly throughout the organization" (p. 138). Marquardt added that, "this transfer of knowledge can occur intentionally and unintentionally. In a learning organization the intentional transfer of knowledge occurs through the following: individually written communication (memos, reports, letters, open access bulletin boards), training, internal conferences, briefings, internal publications, tours, job rotation/transfer, and mentoring" (p. 138). The unintentional learning transfers, however, occur in unplanned interaction among individuals in a learning organization. Marquardt lists the following situations in which the unintentional learning takes place among the individuals in a learning organization: job rotation, stories and myths, task forces, and informal networks.

Organizational learning also has an impact on individual and team learning. In a learning organization, individuals and teams will be able to (Preskill and Torress, 1999, pp. 109–110):

1. Understand how their actions affect other areas of the organization;

2. Tend to ask more questions that give solutions/answers;

3. Develop a greater sense of personal accountability and responsibility for the organizational outcomes;

4. Be more self-directed learners;

5. Take higher risks;

6. Be more consultative, more coaching;

7. Be more likely to ask for help;

8. Be active listeners;

9. Use information to act;

10. Develop creative solutions (willingness to do something different); and

11. Share the works that need to be done.

A learning organization provides a learning culture to individuals and groups/teams. According to Klimecki and Probst (1990), culture is "a system of knowledge and insights which serve as a basis for interpreting experiences and generating actions" (p. 129). Probst (1997) continues to explain that culture is an "implicit phenomenon," and is expressed in shared values, norms, and attitudes among the members of the organization.

However, the organization also faces some challenges when the people inside the organization experience what Probst (1997) called "unlearning." Unlearning is defined by Probst as "the process by which knowledge is erased from the memory" (p. 64). Probst added that "this unlearning process may occur when the organization experiences a defensive pattern, norms and privileges, organizational taboos, and information disorders" (p. 64).

In order to overcome the challenges of learning and sustaining the learning organization, Marquardt (1996, pp. 215–219) listed the following as the solution:

1. Scanning imperative

2. Performance gap

3. Concern for measurement

4. Experimental mindset

5. Climate of openness

6. Continuous education

7. Operational variety

8. Multiple advocates or champions

9. Involved leadership

10. Systems perspective

The organization that is able to transform into and sustain itself as a learning organization will produce the following outcomes (Preskill and Torress, 1999, p. 110):

1. Develop new products and services

2. Increase productivity

3. Have higher morale

4. Improve organizational work climate

5. Experience less turnover

6. Experience less waste/sabotage/error

7. Experience improved financial performance

8. Experience increased efficiency and less redundancy (workers understand how each job contributes to the organization's success)

9. Provide more effective services to clients/customers, and be able to change more quickly

CHAPTER 5

Lesson Learned from the U.S. Concept

Public Health in Indonesia

5.1 PUBLIC HEALTH IN INDONESIA

Public health in Indonesia started to become an integral part of the country's development when the first public health program was initiated in 1956 at three major universities in Indonesia. These programs of public health were under the College of Medicine in three major universities in Jakarta, Yogyakarta and Surabaya. The curriculum of the program was mainly based on the curriculum of public health in the United States as the golden standard.

Ten years after the public health program was introduced through Colleges of Medicine at these three major cities in Indonesia, the Ministry of Education legalized the funding of the College of Public Health in 1965. The Ministry of Education's decree regarding the funding of the College of Public Health in Jakarta was signed on February 26th, 1965, and then there was another decree signed on July 1st, 1965 that substituted that decree.

The aim of the decree for developing the College of Public Health was to provide a sustainable source of trained public health manpower, with the technical knowledge and skills to manage public health and a population service system with emphasis on disease prevention and health promotion. The College of Public Health was also established to create regional public health information centers for the purpose of improving quality of life for the community.

In 1982, the Ministry of Health realized the importance of having human resources in the public health area. The need to have a qualified public health professional encouraged the Ministry of Health to create collaboration with the School of Public Health (SPH) at the University of Hawaii. This collaboration resulted in the establishment of the College of Public Health at five universities in five big cities in Indonesia. In 1985, these Colleges of Public Health opened a 4-year bachelor degree program, and a 3-year diploma program on occupational health and safety.

The early years of the public health program in Indonesia were heavily focused on the medical aspect of health as it is under the College of Medicine. However, as it developed, the public health program spread out its scope to the environmental aspect of health, such as housing, workplace, food and water, community surveillance of health, and community empowerment in emergency responses. The graduates of the Colleges of Public Health dealt with quality assurance, accessibility of health services, high-risk group outreach, policy development and health planning and assessment in their line work. The content of public health program also dealt with communicable and non-communicable diseases, reproductive health, as well as injury prevention and disaster management.

These developments of scope and skill needed for public health professionals contributed to the development of seven disciplines under the College of Public Health in Indonesia. These seven disciplines are health administration and policy, epidemiology, biostatistics and population studies, community nutrition, environmental health, occupational health and safety, as well as health promotion and behavior. These disciplines are needed to help the community get optimum benefit to improve their status.

Health administration and policy aimed to advocate policy makers and health services to create a better service for the community and to enforce a healthy public policy. Epidemiology measures the incidence and prevalence to measure trends in the community to the prevention of particular diseases. Biostatistics and population study help the policy makers to get a complete picture of the demographic condition and create a specific program for a vulnerable population. Community nutrition creates an innovative diet for the community depending on measurement of body mass index to prevent malnourishment. Environmental health evaluates the quality of the environment where people live in order to create a conducive environment for healthy behavior. Health promotion and behavior assess the behavior of the people in the community to identify the main causes of certain behavior and the key person to help promote certain behavior conducive to health.

Public health practitioners believed that in order to help the community improve its health, four activities need to be done: health empowerment, health enhancement, health protection, and health preservation. Health empowerment is conducted to help the community to continue helping itself and others. Health enhancement to improve the community's current health status and to continue improving its health. Health protection means to protect from various hazards in the environment. Health preservation means to continue tradition that contributes to promoting the community's health status.

In order to help the community to improve its status, public health looks at the health problem comprehensively. It considers many factors that influence health and identify stakeholders that contributed to the problem or may be able to help eliminate the problem. This is one of the reasons the public health area incorporates many disciplines such as psychology, communication, policy, anthropology, sociology, and many more depending on the context of issue in public health. These disciplinary fields will also develop depending on the development of issues in public health inside the country or worldwide. In a global world, public health will be influenced by globalization, the condition politics, and the economic aspects of the country and the world.

Public health in Indonesia will be mainly influenced by the existing regulation and the systems of government. These conditions and situations will determine the type of public health problems existing in Indonesia and accessibility to resources to fix those problems. In Indonesia, public health is not only the responsibility of the Ministry of Health, but also the responsibility of other ministries. The health aspect of the school systems, for example, is the responsibility of the Ministry of Education and the Ministry of Health. The multi-stakeholders are supposed to be beneficial for schools, but this can create problems in term of coordination if both ministries lack communication.

In the earlier years, public health in Indonesia focused on the factors that contribute to the health status of the community due to poverty and unsanitary living conditions. These factors include air conditioning, sanitation, water quality, and foodborne diseases. In recent years, public health began looking at the lifestyle of the community (Departemen Kesehatan, 2004). The changes were also influenced by the Alma Ata Declaration in 1978, where it is stated that the main strategy to achieve health for all was through primary health care. In primary health care, the importance of health education was stated as a significant aspect of health care. Health education focused on behavioral changes, but unfortunately has not mentioned the aspect of the environment (physical, biological, and social) that influence the behavioral changes (Departemen Kesehatan, 2004). This is when the concept of health promotion is introduced in the public health program in Indonesia. Health promotion, according to the World Health Organization's Ottawa Charter for Health

Promotion, uses three strategies (enable, mediate, and advocate) to work on five areas of building healthy public policy, creating a conducive environment, developing personal skill, strengthening community action and reorienting health services.

In Indonesia, public health is a broad area that welcomes all practitioners from different fields who are concerned with improving the health status of the community. However, in Indonesia, the term public health experts refers to those professionals who not only are working in the public health area, but also have a bachelor's, master's or doctoral degree in public health. These programs in public health are coordinated under the Ministry of Higher Education, Research and Technology. The ministry also coordinates with a forum that consists of the deans of the schools of public health that have these programs.

5.2 PUBLIC HEALTH PROFESSIONAL COMPETENCIES

Basic areas of competency of the graduates of public health education programs are:

1. Show the application of a combination of knowledge, skill, attitude and behavior

2. Have a critical response to the community and be able to make decisions

3. Be responsible for own decisions and behaviors

4. Provide, facilitate and improve public health services based on the eight main competencies

5. Show a leadership attitude in executing work

6. Pay attention to and have respect for the cultural diversity of the community

7. Provide services to the community

8. Apply a logical and systematic approach in executing work

9. Be prepared to provide wide-scale services (general)

10. Be able to collaborate with other health workers and improve the competencies of each member of the health team

11. Facilitate information-based public health services

12. Understand the nature of public health problems

13. Be proactive in promoting, preventing, and managing risks

14. Increase knowledge and skills through continuing education

These competencies are obtained and measured by the learning outcomes of public health professionals which include their scientific fields, knowledge, and skills. In accordance with the ideology and the culture of Indonesia, the implementations of the national education system carried out at every level of qualification are as follows:

1. Believe in the one supreme God

2. Have a moral, ethical and good attitude in completing duties

3. Act as a proud citizen who loves the homeland and supports world peace

4. Be able to work in a team, have social sensitivity, and be mindful of community and the environment

5. Respect the diversity of cultures, views, beliefs, and religions as well as other people's original opinions/findings

6. Uphold the rule of law and have the spirit to prioritize the interests of the nation and the wider community

According to the demands of essential public health services, as well as understanding more specific problems in the public health education program, education in public health is required to have eight competencies as the basic competencies mastered by public health program graduates. These eight competencies are the result of expert discussions of the Council on Linkages between Academia and Public Health Practice (2001).

In Indonesia, the forum of these deans of colleges of public health is called AIPTKMI (Association of Indonesian Institution of Colleges of Public Health). The members of AIPTKMI meet regularly to discuss the curriculum of the program and how to deliver the courses to achieve competencies of the graduates. Through AIPTKMI, the members develop and make decisions regarding the competencies and sub-competencies of graduates of bachelor's, master's and doctoral programs.

Currently, the competencies of these graduates are the same as and are based on the competencies of public health professionals in the United States. We define competencies as the statement of ability of a graduate from an education or training program that can be measured through a formal exam. As for higher education, competency is defined as the accumulation of someone's ability in conducting a work description that can be measured through structured assessment, including empowerment and personal responsibility in a field of work. The competency of public health professionals has three main functions: assessment, health policy development, and assurance (AIPTKMI, 2012).

The assessment function of the public health professional competency is based on three essential services:

1. Conduct health status monitoring to identify a public health problem

2. Conduct diagnosis and investigation of a health problem and the risk to the community

3. Conduct effectiveness evaluation, accessibility, and assurance of quality services in health care

The health policy development function of the public health professional competency is based on three essential services:

1. Develop policy and program planning that support an individual and public health program

2. Develop policy and regulation to protect and to ensure community health

3. Conduct research to gain new insight and innovative solution for public health

The assurance function of the public health professional competency is based on two essential services:

1. Be able to connect health care for personal need and to ensure health care accessibility

2. Be able to ensure the availability of competent public health professionals

Public health professional competency is comprised of five main elements:

1. Personality traits of public health workers

2. Ability to understand the knowledge, technology and art of public health

3. Ability and skill to become a leader, innovator, advocator, researcher, and manager in planning, implementing and evaluating a health program

4. Have the attitude and behavior in performing according to the mastery of skill

5. Ability to understand the community's lifestyle according to public health principles of ethical and morally accepted values practiced by the local culture

Public health professionals with a bachelor's degree in public health should be able to conduct the following tasks: (IAKMI and AIPTKMI, 2014)

1. Deliver primary health care services for the community by conducting health status monitoring, public health problem diagnosis and investigation

2. Develop and implement operational policy and program planning to support primary health care services

3. Conduct education and community empowerment regarding health and mobilization to identify and solve primary health care problems

4. Conduct monitoring and control primary health care services in terms of effectivity, accessibility, and quality of services

5. Communicate the work to the community and stakeholders of primary health care

6. Understand basic public health knowledge as the requirement to be a manager and worker in primary health care, including basic biomedical information, epidemiology, biostatistics, social and behavioral sciences, environmental health, occupational health and safety, health policy and administration, community nutrition, and reproductive health.

7. Make effective and efficient decisions in technical program planning, organizing, implementing, controlling, assessing and developing alternate solutions to solve public health problems in primary health care

8. Be responsible independently in one's own work, and be critical and responsible for the teamwork

In order to carry out their duties as public health professionals, all graduates of public health education in Indonesia should have eight public health standard competencies as follows.

5.2.1 Ability to Conduct Analysis and Assessment Indicators

Every graduate of public health is capable of collecting, assessing, analyzing and applying the information (including data, facts, concepts and theories) in order to make decisions based on evidence. The objectives of these capabilities are: (a) monitoring health status to identify health problems or dangerous environmental conditions; and (b) diagnosing and investigating health problems by analyzing the conditions of the environment or the behavior of the community which become the health risk factors for disease.

The steps taken are as follows:

1. Define the problem

2. Determine the usefulness and limitations of data

3. Identify appropriate and relevant data and information sources

4. Evaluate data integrity and comparability

5. Use ethical principles in gathering data and information

6. Make relevant inferences from quantitative and qualitative data

7. Take and interpret information related to risks and benefits

8. Implement the process of data collection, IT-based information technology applications

9. Understand how data can clarify the overall issues of public health

5.2.2 Ability to Plan and Develop Health Policies (Policy Development and Program Planning)

Every graduate of public health is capable of making a decision, planning, implementing, and evaluating based on the evidence, including managing the outbreaks such as plagues and emergencies. Steps include:

1. Gather, summarize and interpret information about health issues

2. State the policy and write it clearly and in detail

3. Address health, fiscal, administrative, legal, social and political implications

4. State the expected feasibility and outcome in every policy

5. Use the latest techniques in the determination and planning analysis of health

6. Decide on the appropriate actions

7. Develop a plan to implement policies

8. Change policies into organizational, structural and program plans

9. Prepare and implement an emergency response plan

10. Develop a program monitoring and evaluation mechanism

5.2.3 Ability to Communicate (Communication Skills)

Every graduate of public health has the ability to interact and communicate internally and externally, verbally and non-verbally, and by listening; use a computer; provide the right information for different audiences; and work with media and social marketing.

The steps to good communication are as follows:

1. Gather, summarize and interpret information about health issues

2. State policy choices and write clearly and densely

3. Address health implications: fiscal, administrative, legal, social and political

4. State the expected feasibility and outcome of each policy choice

5. Use the latest techniques in health determination and planning analysis

6. Decide on appropriate actions

7. Develop a plan to implement policies

8. Change policies into organizational, structural and program plans

9. Prepare and implement the plan of emergency response

10. Develop the mechanism of program monitoring and evaluation

5.2.4 Ability to Understand Local Culture (Cultural Competency/Local Wisdom)

Every graduate of public health is capable of identifying and understanding the socio-cultural competencies needed to build effective interactions with a variety of individuals, groups and communities, as a manifestation of moral and ethical attitudes towards socio-cultural factors for successful implementation of behavior, programs, and policies.

The steps to cultural competency are as follows:

1. Use the right methods to interact sensitively, effectively and professionally with people of different cultural, socioeconomic, educational, racial, ethnic and professional backgrounds, in all age groups and their lifestyles

2. Use appropriate methods to interact sensitively, effectively and professionally with people of different cultural, socioeconomic, educational, racial, ethnic and professional backgrounds, in all age groups and their lifestyles

3. Develop and adapt approaches to problems related to cultural differences

4. Understand the dynamics that contribute to cultural diversity (attitude)

5. Understand the importance of diversity on public health workers (attitude)

5.2.5 Ability to Empower the Community (Community Dimensions of Practice)

Every graduate of public health is capable of empowering the community. They are skilled in empowering the potency of community by using community development and organizing methods. Moreover, they also should advocate, work, and partner with the community to achieve mutual goals.

The steps to empowering the community are as follows:

1. Able to combine various strategies to interact with people from various backgrounds

2. Identify the role of social, cultural, and behavioral factors in health

3. Respond to various needs as a consequence of cultural diversity

4. Identify and maintain relationships with stakeholders and community leaders

5. Use group dynamics processes to enhance community participation

6. Describe the role of the government in providing social security

7. Describe the role of the government and the private sector in providing public health services

8. Identify the potencies and resources of the community

9. Gather input from the community as information/material for consideration in the development of Public health policies and programs

10. Relay program policies and resources to inform the community

5.2.6 Understand the Basics of Public Health Sciences

Every graduate of public health understands the main knowledge of public health, the critical thinking skills that are related to the disciplines of the public health sciences.

The steps to understanding the basics of public health sciences are as follows:

1. Identify the obligations of individuals and organizations in the context of essential public health services and the basic functions

2. Define, assess, and understand the health status of the population, determinants of health and diseases, factors that contribute to health promotion and prevention of disease

3. Understand the historical development, structure, and interaction between the public health and health service systems

4. Identify and apply basic research methods used in public health

5. Use group dynamics processes to enhance community participation

6. Apply the public health science including social sciences and behavior of chronic disease, infection, and accidents

7. Identify the limitations of research and the importance of observation and interrelationships

8. Develop an all-time commitment to strong critical thinking (attitude)

5.2.7 Ability to Plan and Manage Funding Sources (Financial Planning and Management)

Graduates of public health are capable of developing and presenting budgets based on program direction for budget efficiency and effectiveness, as well as cost utilization.

The steps to planning and managing funding are as follows:

1. Develop and present a budget

2. Manage programs with a limited budget

3. Implement the budget process

4. Develop strategies to determine budget priorities

5. Monitor program performance

6. Prepare proposals to obtain funds from external sources

7. Implement basic human relations skills in organizational management, staff motivation, and conflict resolution

8. Negotiate and develop contracts and other documents for the provision of community-based services

9. Make an analysis of cost effectiveness, cost benefit, and cost utility

5.2.8 Ability to Lead and Think Systems (Leadership and Systems Thinking/Total System)

Graduates of public health are able to lead and build capacity, improve performance, improve the quality of the work environment, and have leadership skills to stimulate organizations and the community to create and implement vision, mission and values, in the aim of improving the level of public health.

The steps to leadership and systems thinking are as follows:

1. Create a culture of standard ethics within organizations and communities

2. Create basic values and shared visions, and use these principles in the implementation guidelines

3. Identify internal and external issues which are capable to impact the implementation of essential public health services (e.g., strategic plans)

4. Facilitate the collaboration of internal and external groups to ensure the participation of key stakeholders

5. Contribute to the development, implementation and monitoring of the organizational work standards

6. Use the legal and political system to make changes

7. Apply the theory of organizational structure to professional practice

5.3 CURRICULUM

In order to prepare professional public health workers for their respective careers, the role of the university is very important as an institution that creates professional candidates. Through education at the university level, students are exposed to various scientific and public health practices that contribute to the achievement of their competencies.

Academic education is a system of higher education directed at mastering and developing certain disciplines of science, technology, and art, which include bachelor's, master's, and doctoral programs.

Generally, a public health bachelor's program is directed to the application of public health science, a master's program is directed to the development of public health science, and a doctoral program is directed to discovery of more findings in public health science. A public health educational program also provides a professional program which prepares students to master special skills after attending their bachelor's program.

A public health bachelor's program is an undergraduate program with eight main competencies, with the basic principles of Kerangka Kualifikasi Nasional Indonesia (KKNI), that teach the ability to understand the basic principles of public health, to implement skills, and to lead and work together as a team. Supporting competencies and other competencies are developed to meet the needs of specialization, specific characteristics of higher education based on local needs. The bachelor's program requires a minimum of 144 credits and a maximum of 160 (Ministry of National Education decree number 232 of 2000) to fulfill the eight main competencies, supporting competencies, and other competencies as a candidate.

The standard curriculum for the public health bachelor's program requires the completion of 144 credits, with a composition consisting of:

a. Compulsory courses (70%) in:

1. National public health bachelor's study program (86 SKS or credit points)

2. Local content study program (15 SKS or credit points)

b. Compulsory courses in a specific major (30%)

The total credits for a bachelor's program in public health is 146 credits. Table 5.1 describes the percentage of each semester of these total credits.

5.4 COMPETENCY TEST

There was a need to implement the Internal Quality Assurance System in the institutions that administer bachelor's programs,

TABLE 5.1   Percentage of Credits per Semester for a Bachelor's Program in
Public Health

Semester	Major	Credits for Each Major			Total Credits
		Lecture	Tutorial	Practicum	
1	–	20	0	0	20
2	–	20	0	0	20
3	–	16	0	0	16
4	–	17	0	2	19
5	–	18	0	3	21
6	Health Policy and Administration	23	0	0	23
	Biostatistics	24	0	0	24
	Reproductive Health	23	0	0	23
	Epidemiology	24	0	0	24
	Nutrition	20	0	2	22
	Occupational Health and Safety	21	0	0	21
	Environmental Health	19	0	3	22
	Health Promotion and Behavior	19	0	3	22
7	Health Policy and Administration	17	0	3	20
	Biostatistics	12	0	7	19
	Reproductive Health	15	0	5	20
	Epidemiology	14	0	5	19
	Nutrition	18	0	3	21
	Occupational Health and Safety	17	0	5	22
	Environmental Health	18	0	3	21
	Health Promotion and Behavior	18	0	3	21
8	Health Policy and Administration	138	0	8	146
	Biostatistics	134	0	12	146
	Reproductive Health	136	0	10	146
	Epidemiology	136	0	10	146
	Nutrition	136	0	10	146
	Occupational Health and Safety	136	0	10	146
	Environmental Health	135	0	11	146
	Health Promotion and Behavior	135	0	11	146

especially in terms of standardization and guaranteeing the quality of graduates. Therefore, it was necessary to implement a quality competency test as part of the learning evaluation process integrated in the educational system. A competency test is a process to measure the knowledge, skills and attitudes of health workers in accordance with professional standards.

The Indonesian Bachelor of Public Health Competency Test was developed and held as fulfillment of the mandate of the Republic of Indonesia Law No. 12 of 2012 concerning higher education, which was then followed by the Minister of Education and Culture Regulation No. 83 of 2013 concerning competency certificates, and Joint Regulations of the Minister of Health and the Minister of Education and Culture No. 36 of 2013 and No. 1/IV/PB/2013 concerning competency tests for higher education students of health.

The public health graduates' competencies ensure that a person is capable of carrying out tasks in certain health occupations. A graduate competency test is a process to measure the knowledge, skills, and attitudes of public health graduates in accordance with the required competencies. Moreover, to guarantee the implementation of a qualified test, the blueprint of the Public Health Bachelor Competency Test is prepared as a guidance for making up questions and executing the competency test.

The aims of conducting a national competency test for new graduates are as follows:

1. To ensure that public health higher education graduates are competent and meet minimum national standard of profession

2. To test the knowledge and skills as a basis for creating professionalism in service and encouraging lifelong learning

3. To uphold the professional accountability of public health education graduates in carrying out their professional roles

4. To assess management of safe and effective health services and protecting public trust in public health graduate professions

5. To improve the quality of higher education in public health on an ongoing basis

The competency test for graduates of public health is carried out by using the paper-based test method. The computer-based test method and Internet-based test method are also used after making sure that every corner of the region has appropriate facilities to be used as a place for conducting competency tests. Determination of the method used will be carried out by the central organizer in accordance with the feasibility of the place of examination.

A paper-based test is a test method where the questions are recorded in the form of booklets containing rules and questions. Answer sheets are made and given separately. The booklet has a security seal that tears if it is opened. The integrity of the security seal is considered as a guarantee of confidentiality. The questions can be answered by filling in the circles with 2B pencils according to the answer choices.

Computer-based and Internet-based tests are test methods that use computers and Internet networks. The questions are prepared online or on portable hard drives. Participants are given a login account. Participants' answers are stored online or stored on a hard disk which is then sealed and submitted to the central organizer.

Passing grade are set together by a team formed by a panel of experts consisting of through discussion and analysis of the difficulty level of the questions. The agreed grading method is the "modified Angoff method" which guarantees that the passing grade is valid and fair. The use of the *Angoff Method* is carried out as follows:

a. Panelists look at the first question individually and assess the level of difficulty

b. Each panelist individually estimates the percentage of a group of test takers who can answer the question correctly

c. Panelists discuss their estimation results

d. Each estimation result is tabulated and the averages are calculated

e. The above sequence is repeated for all the questions

f. The average estimation result for each question is added up and averaged to get the cut point

The number of questions used in the competency test is 180 to be answered in 180 minutes. The types of questions are multiple-choice questions (MCQ type A question) with 5 alternative answers (A, B, C, D, and E) where one chooses the best answer. The number of questions is considered capable of measuring a new graduate's competency accurately (fulfilling the reliability of the question). The questions also have gone through the process of validity testing. Provisions for implementing the test are adjusted according to the guidelines for the implementation of the test.

Each question consists of a vignette (case), lead-in (question) and options (answer choices). A vignette describes a situation that is presented in a focused, logical and systematic manner. One vignette can be used for a maximum of two questions. The questions are presented in the form of the "recall" and reasoning types of questions that are capable of measuring cognitive, psychomotor or affective skills.

Rules used in making up questions are:

a. Focus of questions. Questions are geared toward the 8 areas of public health competency that later are measured by method review and that focus on the public health field, including biostatistics and populations, policy administrations, health and behavioral education, occupational health, reproduction health and community nutrition. Questions are made to portray the situation within the intervention target: family, community, and health service institution, either in primary, secondary, or tertiary health service hierarchy.

b. Analyzing argumentation. When given the situation, participants are able to provide reasons that support the existing argumentation.

c. Making conclusions. When given a statement, participants are able to conclude in accordance to the statement.

d. Assessing. When given the problem statement, participants are able to solve the problem served through appropriate reasons.

e. Describing a concept or an assumption. When given an argument, participants are able to determine the correct theory or assumption.

f. Describing the field situation. When given the situation on field, participants are able to describe statement or field data that has been fixed, so that the representation of intact situation can be fully obtained.

g. Solving problem systematically. When given a statement, participants are able to solve problems systematically using scientific rules.

h. Evaluating strategy. When given the statement of problem or strategy, participants are able to evaluate the strategy or procedure available and can choose the efficient and effective procedure.

5.5 PROFESSIONAL PUBLIC HEALTH ETHICS

The public health profession should be able to convince society of its integrity. In addition, society should feel confident that the public health profession is able to provide solutions, advice, and steps to prevent and control disease through promoting and preventing according to the needs, ethics, and norms that lie within society. The most important thing that the public health profession is able to obtain from society is trust. Therefore, the public health profession is able to be trusted and able to defend what it believes for a purpose of realizing social justice and increasing public health standards.

In Indonesia, there is a discourse that public health graduates will be given choices to continue obtaining professional degrees after graduation from a public health program. This professional education will be compulsory in the form of a 1-year program after finishing an undergraduate program. At this time, it remains a discussion about whether public health educational programs will become a general profession or a concentrated continuation. Even though currently there are no professional ethics schools, public health graduates should take a vow to uphold their work ethics. Ethical attitude and behavior later will be benchmarks for public health professional ethics.

The word "ethics" derives from the Greek, Ethos, which means "custom," "behavioral model," or expected standard and particular criteria for some attitudes. The term is currently interpreted as a motive to encourage changed behavior. Ethics is a behavioral code that indicates a good deed for particular groups. Ethics also stands for rules and principles for a good deed and the connection between good or bad things and moral obligation. It also relates to rules for attitude or deed, which contain a right or wrong principle, as well as a morality principle, since ethics holds a moral responsibility. This means that a deviation from ethical code can be interpreted as doing a bad deed and having bad morals.

Ethics is the sense of value or morals that becomes a guidance for a person or group to set behavior that can be likened to customs or habits. This means that ethics can be defined as a compilation of principles of moral value that are also known as ethical code. Ethics can turn into science if the accepted ethical possibilities become reflection materials for some systematic and methodical research.

Public health ethical principles are as follows:

1. Public health mainly should discuss the basic cause of disease and requirement to prevent harm to health

2. Public health should respect individual rights that apply within society

3. Public health policy, programs, and priority should be developed and evaluated through a process that ensures opportunity of suggestion from society members

4. Public health should advocate or work for society empowerment to make sure that basic resources and conditions necessary to achieve health can be accessed by all people in society

5. Public health should seek information necessary for implementing effective policy and programs that cover and improve health

6. Public health institutions should provide information that they possess for decisions about policies or programs that require society's consent for their implementation

7. Public health institutions should act on time based on the update information and based on the need of the community.

8. Public health programs and policy should merge various approaches that anticipate and respect diverse values, beliefs, and cultures that lie within society

9. Public health programs and policy should be undertaken in the best way to enhance the physical and social environment

10. Public health institutions should protect information confidentiality that can potentially harm individuals or communities if published. Exceptions only can be justified based on high possibility to harm individuals.

11. Public health institutions should guarantee their employees' professional competency.

12. Public health institutions and their employees should collaborate and connect in a way that can build public trust and institution affectivity.

Ethical dilemma occurs when a public health problem requires reporting to the health authority in charge. The report does not infringe on ethics if the risk to the public has the following features: high risk probability, serious risk in effect, and risk related to individual or group is possible to identify. Currently, there is a development of the Public Health Ethical Code, which is elaborated on in the chapters and articles below (IAKMI, 2013):

1. General Responsibility (Chapter 1) consists of 5 articles (Articles 1–5):

   Article 1: Every public health professional should honor, embrace, and implement public health profession ethics.

   Article 2: While doing their job, public health professionals should prioritize public interest more than personal interest.

   Article 3: While carrying out their job and responsibilities, public health professionals should implement efficiency and affectivity principles and prioritize appropriate technology usage.

   Article 4: While carrying out their job and responsibilities, public health professionals should not be classifying the public based on religion, ethnicity, race, socio-political, or any other consideration.

   Article 5: While doing their job and function, public health professionals merely perform their own profession and skill.

2. Responsibility to Society (Chapter 2) consists of 8 articles (Articles 6–13):

Article 6: While doing their job and function, public health graduates should always be society-oriented as a unit that is related to social, economic, political, psychological, and cultural aspects.

Article 7: While doing their job and function, public health graduates should put community's health maintenance first.

Article 8: While doing their job and function, public health graduates should put justice and equity first.

Article 9: While doing their job and function, public health graduates should apply a holistic approach, multi-disciplinary and cross-sector, and prioritize to promote, prevent, protect, and develop health maintenance.

Article 10: The health maintenance effort should be based on scientific facts that are obtained by literature reviews or research.

Article 11: While maintaining public health, publick health graduates should use professional procedures and steps that have been tested through scientific study.

Article 12: While doing their job and function, public health graduates should be responsible for protecting, preserving, and improving population health.

Article 13: While doing their job and function, public health graduates should base on forward anticipation, either related to health problem or another problem that influences population health.

3. Responsibility to Another Health Profession and Outside Health Sector Profession (Chapter 3) consists of 2 articles (Articles 14 and 15):

Article 14: While doing their job and function, public health graduates should cooperate and respect each other regardless of their belief, religion, ethnic, race, etc.

Article 15: While doing their job and function along with another professional, public health graduates should adhere to such principles as: partnership, leadership, initiation making and pioneering.

4. Responsibility to Profession (Chapter 4) consists of 2 articles (Articles 16 and 17):

   Article 16: Public health professionals should be proactive in overcoming situations.

   Article 17: Public health professionals should be constantly maintaining and enhancing the public health profession.

   Article 18: Public health professionals should be constantly communicating and sharing experiences, and helping each other.

5. Responsibility to Self (Chapter 5) consists of 2 articles (Articles 19 and 20):

   Article 19: Public health professionals should maintain their own health in order to do their job and profession well.

   Article 20: Public health professionals should constantly attempt to enhance knowledge and skill related to science and technology.

6. Commitment to Obey Ethical Professional Conduct (Chapter 6) consists of 1 article (Article 21):

   Article 21: Every member of the public health professional association in conducting their responsibilities must hold the ethical conduct of public health professionals in this document.

Bibliography

Abrahamson, S. 1985. *Evaluation of Continuing Education in the Health Professions.* (S. Abrahamson, Ed.). Boston: Kluwer Academic Publishers.

Academia and Public Health Practice. 2001. *Core Competencies for Public Health Professionals.* Washington, DC: Council on Linkages Between Academia and Public Health Practice.

Allegrante, J. P., Moon, R. W., Auld, M. E., and Gebbie, K. M. 2001. Continuing-education needs of the currently employed public health education workforce. *American Journal of Public Health,* 91(8), 1230–1234.

American Public Health Association (APHA) 2019a. CE Mission and Accreditations. Retrieved May 5, 2013, from https://www.apha.org/professional-development/continuing-education/ce-mission-and-accreditations

American Public Health Association (APHA) 2019b. Continuing education accreditation. Retrieved May 5, 2013, from https://www.apha.org/professional-development/continuing-education/continuing-education-accreditation

Anderson, M. B. 1999. Public health in medical education: Where are we now? In: *Education for More Synergistic Practice of Medicine and Public Health.* New York: Josiah Macy, Jr., Foundation, 195–205.

Association of Indonesian Public Health Higher Education Institutions (AIPTKMI). 2012. *Academic Script of Public Health Education.* Jakarta: Aiptkmi.

ASPH (Association of Schools of Public Health)/Special Study Committee. 1966. *The Role of Schools of Public Health in Relation to Trends in Medical Care Programs in the United States and Canada.* Alan Mason Chesney Archives of the Johns Hopkins Medical Institutions.

ASPH. 2000. *1999 Annual Data Report: Applications, New Enrollments, and Students; Fall 1999 Graduates; 1998-1999 with Trends, 1989-1999.* Washington, DC: ASPH.

Beckett, D. 2001. Hot action at work: A different understanding of "understanding." *New Directions for Adult and Continuing Education*, 92, 73–84.

Benner, P. E. 2010. *Educating Nurses: A Call for Radical Transformation.* San Francisco: Jossey-Bass.

Bennis, W. G. and Shepard, H. A. 1956. A theory of group development. *Human Relations*, 9(4), 415–437. https://doi.org/10.1177/001872675600900403.

Beyle, T. and Dusenbury, P. 1982. Health and human services block grants: The state and local dimension. *State Government*, 55(1), 2–13.

Bialek, R. G. 2001. *Council on Linkages between Academia and Public Health Practice: Bridging the Gap Progress Report, July 1 through September 30.* Washington, DC: Public Health Foundation.

Blockstein, A. M. 1977. *Graduate School of Public Health, University of Pittsburgh, 1948-1974.* Pittsburgh: University of Pittsburgh.

Bower, E. A., Choi, D., Becker, T. M., and Girard, D. E. 2007. Awareness of and participation in maintenance of professional certification: A prospective study. *The Journal of Continuing Education in the Health Professions*, 27(3), 164–172.

Carlson, V., Chilton, M. J., Carso, L. C., and Beitsch, L. M. 2015. Defining the functions of public health governance. *American Journal of Public Health*, 105(2), 159–166. https://doi.org/10.2105/AJPH.2014.302198

CDC. 2001b. [Online]. Available: www.cdc.gov/PHTN/history.htm

CDC. 2001c. [Online]. Available: www.phppo.cdc.gov/workforce

Centers for Disease Control and Prevention (CDC). 2018. The Public Health System and the 10 Essential Public Health Services. Retrieved March 23, 2019, from https://www.cdc.gov/publichealthgateway/publichealthservices/essentialhealthservices.html

Center for Health Policy. 2000. *The Public Health Workforce: Enumeration 2000.* New York: Columbia University School of Nursing.

Cervero, R. M. 1988. *Effective Continuing Education for Professionals.* (R. M. Cervero, Ed.). San Francisco: Jossey-Bass.

Cervero, R. M. 2003. Place matters in physician practice and learning. *The Journal of Continuing Education in the Health Professions*, 23(Suppl 1), S10–S18.

Champoux, J. E. 2010. *Organizational Behavior: Integrating Individuals, Groups, and Organizations* (4th ed.). London: Routledge Taylor & Francis Group.

Chen, L.-S. and Goodson, P. 2007. Public health genomics knowledge and attitudes: A survey of public health educators in the United States. *Genetics in Medicine: Official Journal of The American College of Medical Genetics*, 9(8), 496–503.

Committee on Professional Education. 1937. Public health degrees and certificates granted in 1936. *American Journal of Public Health*, 27, 1267–1272.

Conrad, D. 2000. Bringing two worlds closer: A three-year review of council activities. *The Link*, 14(1), 1–4.

Cope, P., Cuthbertson, P., and Stoddart, B. 2000. Situated Learning in the practice placement. *Journal of Advance Nursing*, 31(4), 850–856.

Council on Linkages. 2001. *Core Competencies for Public Health Professionals*. Washington, DC: Public Health Foundation.

Corvey, R. J. 2003. The role of listserv participation in the professional development of a nursing community of practice. Retrieved May 5, 2013, from http://purl.galileo.usg.edu/uga_etd/corvey_rebecca_j_200312_edd

Council of State Governments. 1987. *The Book of the States, 1986–87* (vol. 26). Lexington, KY.

Davidson, E. S. 2008. Perceived continuing education needs and job relevance of health education competencies among health education and promotion practitioners in college health settings. *Journal of American College Health*, 57(2), 197–210.

Davis, G. B. 2002. *Basic Framework: Management Information Systems, Part I Introduction*. Management Series No. 90-A. Twelfth Printing. Jakarta: PT. Binawan Pressindo Library.

Davis, M. V. and Dandoy, S. 2001. *Survey of Graduate Programs in Public Health and Preventive Medicine and Community Health Education*. Final Report supported by ATPM/HRSA cooperative agreement 6U76AH000001. Washington, DC.

Demers, A. R. and Mamary, E. 2008. Assessing the professional development needs of public health educators in light of changing competencies. *Preventing Chronic Disease*, 5(4), A129.

Dennis, D. L. and Lysoby, L. 2010. The advanced credential for health education specialists: A seven-year project. *Health Educator*, 42(2), 77–83.

Denton, J. 1998. *Organisational Learning and Effectiveness*. London: Routledge.

Department of Health Commonwealth of Virginia. 1984. *The Health Laws of Virginia*. Charlottesville, VA: The Mitchie Co.

Departemen Kesehatan (Depkes), RI. 2004. National Health System 2004, Jakarta.

Desikan, N. 2009. *"Communities of Practice for Continuing Professional Education [Electronic Resource]: A Case Study of Educational Consultants in India."* PhD diss., University of Georgia.

DiClemente, R. J., Salazar, L. F., and Crosby, R. A. 2013. *Health Behavior Theory for Public Health: Principles, Foundations, and Applications*. Burlington, MA: Jones & Bartlett Learning.

Dirani, K. M. 2013. Does theory travel?: Dimensions of the learning organization culture relevant to the Lebanese culture. *Advances in Developing Human Resources*, 15(2), 177–192. https://doi.org/10.1177/1523422313475992

Dixon, N. M. 1999. *The Organizational Learning Cycle: How We Can Learn Collectively* (2nd ed.). Brookfield, VT: Gower.

Eisen, K., Flake, M., and Wojciak, A. 1994. *How, Where and When does Theory Meet Practice? The Link*. Baltimore, MD: Johns Hopkins University Health Program Alliance.

Ellery, J., Allegrante, J. P., Moon, R. W., Auld, M. E., and Gebbie, K. 2002. Continuing-education needs of the currently employed public health education workforce. *American Journal of Public Health*, 92(7), 1053–1054. https://doi.org/10.2105/ajph.92.7.1053

Evans, P. P. 2002. *An accreditation perspective on the future of professional public health preparation*. Presentation to the Institute of Medicine Committee on Educating Public Health Professionals for the 21st Century, Irvine, CA.

Falk, J. K. and Drayton, B. 2009. *Creating and Sustaining Online Professional Learning Communities*. New York: Teachers College Press.

Fee, E. and Rosenkrantz, B. 1991. Professional education for public health in the United States. *A History of Education in Public Health: Health that Mocks the Doctors' Rules*. E. Fee and R. M. Acheson (Eds.). Oxford, UK: Oxford University Press, 230–271.

Fineberg, H. V., Green, M., Ware, J. H., and Anderson, B. L. 1994. Changing public health training needs: Professional education and the paradigm of public health. *Annual Review of Public Health*, 15, 237–257.

Finocchio, L. J., Love, M. B., and Sanchez, E. V. 2003. Illuminating the MPH health educator workforce: Results and implications of an employer survey. *Health Education & Behavior*, 30(6), 683–694. https://doi.org/10.1177/1090198103255365

Gabbay, J., le May, A., Jefferson, H., Webb, D., Lovelock, R., Powell, J., and Lathlean, J. 2003. A case study of knowledge management in multi-agency consumer-informed "communities of practice": Implications for evidence-based policy development in health and social services. *Health: An Interdisciplinary Journal for the Social Study of Health, Illness & Medicine,* 7(3), 283–310.

Gebbie, K. M. 1999. The public health workforce: Key to public health infrastructure. *American Journal of Public Health,* 89(5), 660–661.

Gebbie, K., Rosenstock, L., and Hernandez, L. M. (Eds.). 2003a. *History and Current Status of Public Health Education in the United States.* Washington, DC: National Academies Press. Retrieved from: https://www.ncbi.nlm.nih.gov/books/NBK221176/

Gebbie, K., Rosenstock, L., and Hernandez, L. 2003b. Who Will Keep the Public Healthy? Educating Public Health Professionals for the 21st Century. In *Committee on Educating Public Health Professionals for the 21st Century.* Washington, DC: National Academy Press.

George, D. R. 2011. "Friending Facebook?" A minicourse on the use of social media by health professionals. *Journal of Continuing Education in the Health Professions,* 31(3), 215–219. https://doi.org/10.1002/chp.20129

George, J. M. and Jones, G. R. 2002. *Understanding and Managing Organizational Behavior.* New Jersey: Prentice Hall.

Gill, S. J. 2010. *Developing a Learning Culture in Nonprofit Organizations.* Los Angeles: SAGE.

Gilmore, G. D., Olsen, L. K., Taub, A., and Connell, D. 2005. Overview of the national health educator competencies update project, 1998–2004. *Health Education & Behavior,* 32(6), 725–737.

Ginzberg, E. and Dutka, A. B. 1989. *The Financing of Biomedical Research.* Baltimore: The Johns Hopkins Press.

Glascoff, M. A., Johnson, H. H., Glascoff, W. J., Lovelace, K., and Bibeau, D. L. 2005. A profile of public health educators in North Carolina's local health departments. *Journal of Public Health Management & Practice,* 11(6), 528–536.

Gordon, L. J. and McFarlane, D. R. 1996. Public health practitioner incubation plight: Following the money trail. *Journal of Public Health and Policy,* 17(1), 59–70.

Grayston, J. T. 1974. New approaches in schools of public health: The University of Washington School of Public Health and Community Medicine. *Schools of Public Health: Present and Future, A Report of a Macy Conference.* J. Z. Bowers and E. F. Purcell (Eds.). New York: Josiah Macy, Jr., Foundation, 49–59.

Grad, F. P. 1981. *Public Health Law Manual: A Handbook on the Legal Aspects of Public Health Administration and Enforcement.* Washington, DC: American Public Health Association.

Green, L., Daniel, M., and Novick, L. 2001. Partnerships and coalitions for community-based research. *Public Health Reports (Washington, DC: 1974)*, 116(Suppl 1), 20–31. doi: 10.1093/phr/116.S1.20

Guskey, T. R. 2000. *Evaluating Professional Development.* Thousand Oaks, CA: Corwin Press.

Haid, M., Schroeder, D., Sims, J., and Wang, H. 2010. *Organization Effectiveness.* Philadelphia: Right Management.

Hanlon, J. J. and Pickett, G. E. 1984. *Public Health Administration and Practice.* (8th ed). Saint Louis: Times Mirror/Mosby.

Halverson, P. K., Mays, G., Kaluzny, A. D., and House, R. M. 1997. Developing leaders in public health: The role of the executive training programs. *Journal of Health Administration Education*, 15(2), 87–100.

Harris Jr, J. M. 2009. Information seeking in the digital age—Why closing knowledge gaps is not education and why the difference matters. *Journal of Continuing Education in the Health Professions*, 29(4), 276–277. https://doi.org/10.1002/chp.20047

Hirotsugu, A. 2006. Reasons for participation in and needs for continuing professional education among health workers in Ghana. *Health Policy*, 77(3), 290–303. https://doi.org/10.1016/j.healthpol.2005.07.023

Ho, K., Jarvis-Selinger, S., Norman, C. D., Li, L. C., Olatunbosun, T., Cressman, C., and Nguyen, A. 2010. Electronic communities of practice: Guidelines from a project. [Article]. *Journal of Continuing Education in the Health Professions*, 30(2), 139–143.

Houle, C. O. 1980. *Continuing Learning in the Professions.* (C. O. Houle, Ed.). (1st ed.): San Francisco, CA: Jossey-Bass Publishers.

IAKMI and AIPTKMI. 2014. *Blue Print Uji Kompetensi Sarjana Kesehatan Masyarakat Indonesia.* Jakarta: DKI.

Ikatan Ahli Kesehatan Masyarakat Indonesia. 2013. *Kode Etik Profesi Kesehatan Masyarakat Indonesia.* Jakarta: PP IAKMI.

Institute of Medicine, Division of Health Care Services, Committee for the Study of the Future of Public Health. 1988. *The Future of Public Health.* Washington, DC: National Academy Press. Retrieved from https://books.google.co.id/books?id=pHDt0C3LEagC&pg=PA173&lpg=PA173&dq=American+Medical+Association,+Dep artment+of+State+Legislation,+1984&source=bl&ots=Jh2k24 bxcP&sig=ACfU3U3lfQePyBBNLxND_0_UpRBIpY450g&hl=en &sa=X&ved=2ahUKEwjLvqHej5jhAhXEfn0KHaD7DSoQ6AEw

IOM (Institute of Medicine). 1988. *The Future of Public Health.* Washington, DC: National Academy Press.

Jarvis, P. 1985. *The Sociology of Adult and Continuing Education.* Beckenham: Croom Helm.

Jewell, L. and Siegall, M. 1992. *Contemporary Industrial/Organizational Psychology.* Jakarta: Arcan.

Jewell, L. N. and Siegall, M. 1990. *Modern Organizational Industrial Psychology.* Danuyasa (Ed.). (2nd ed.). Jakarta: Arcan.

Johnson, H. H., Glascoff, M. A., Lovelace, K., Bibeau, D. L., and Tyler, E. T. 2005. Assessment of public health educator practice: Health educator responsibilities. *Health Promotion Practice,* 6(1), 89–96.

Joint Committee on Health Education and Promotion Terminology. 2002. Report of the 2000 joint committee on health education and promotion terminology. *Journal of School Health,* 72(1), 3–7.

Kasworm, C. E., Rose, A. D., and Ross-Gordon, J. M. 2010. *Handbook of Adult and Continuing Education.* (C. E. Kasworm, A. D. Rose, J. M., and Ross-Gordon, Eds.) (2010th ed.). Los Angeles: SAGE.

Kim, Y.-S. and Marsick, V. J. 2013. Using the DLOQ to support learning in Republic of Korea SMEs. *Advances in Developing Human Resources,* 15(2), 207–221. https://doi.org/10.1177/1523422313475994

Klimecki, R. G. and Probst, G. J. B. 1990. *Origin and Development of the Corporate Culture.* Lattmann (Ed.). Heidelberg: The Corporate Culture, 41–67.

Knowles, M. et al. 1984. *Andragogy in Action. Applying Modern Principles of Adult Education.* San Francisco: Jossey Bass.

Knowles, M. S. 1973; 1990. *The Adult Learner. A Neglected Species.* (4th ed.). Houston: Gulf Publishing. 2e. 292 + viii pages. Surveys learning theory, andragogy and human resource development (HRD).

Lathlean, J. and May, A. 2002. Communities of practice: An opportunity for interagency working. *Journal of Clinical Nursing,* 11(3), 394–398. Retrieved from https://europepmc.org/abstract/med/12010537

Lindsay, L. N. 2000. *"Transformation of Learners in a Community of Practice Occupational Therapy Fieldwork Environment."* PhD diss., University of Georgia.

Marquardt, M. J. 1996. *Building the Learning Organization: A Systems Approach to Quantum Improvement and Global Success.* New York: McGraw Hill.

Marsick, V. J. 2013. The dimensions of a learning organization questionnaire (DLOQ): Introduction to the special issue examining DLOQ use over a decade. *Advances in Developing Human Resources,* 15(2), 127–132. https://doi.org/10.1177/1523422313475984

■ Bibliography

Marsick, V. J. and Watkins, K. E. 1990. *Informal and Incidental Learning in the Workplace*. London: Routledge.

Marsick, V. J. and Watkins, K. E. 2003. Demonstrating the value of an organization's learning culture: The dimensions of the learning organization questionnaire. *Advances in Developing Human Resources*, 5(2), 132.

Mattheos, N., Schittek, M., Attstrom, A., and Lyon, H. C. 2001. Distance learning in academic health education. *European Journal of Dental Education*, 5, 67–76.

McNeil, J. R. 2000. *Something New Under the Sun: An Environmental History of the Twentieth Century World*. New York: W.W. Norton, 339.

Merriam, S. B. 2001. Book reviews. *Adult Education Quarterly*, 51(4), 344–345. https://doi.org/10.1177/074171360105100408

Merriam, S. B. and Caffarella, R. S. 1991. *Learning in Adulthood. A Comprehensive Guide*. San Francisco: Jossey-Bass.

MMF (Milbank Memorial Fund). 1976. *Higher Education for Public Health: A Report of the Milbank Memorial Fund Commission*. New York: Prodist.

NASPAA (National Association of Schools of Public Affairs and Administration). 2002. [Online]. Available: http://www.naspaa.org/sur98_4.htm

National Commission for Health Education Credentialing. 2006. *A Competency-Based Framework for Professional Development of Certified Health Education Specialists*.

National Commission for Health Education Credentialing Inc. 2013. Responsibilities and Competencies for Health Education Specialists. Retrieved March 23, 2019, from https://www.nchec.org/

Oliver, T. R. 2006. The politics of public health policy. *Annual Review of Public Health*, 27(1), 195–233. doi:10.1146/annurev.publhealth.25.101802.123126

Olson, C. A. 2012. Focused search and retrieval: The impact of technology on our brains. *Journal of Continuing Education in the Health Professions*, 32(1), 1–3.

Omenn, G. S. 1982. "What's behind those block grants in health?" *New England Journal of Medicine*, 306(17), 1057–1060.

Pereles, L., Lockyer, J., and Fidler, H. 2002. Permanent small groups: Group dynamics, learning, and change. *The Journal of Continuing Education in the Health Professions*, 22(4), 205–213.

PHLS (Public Health Leadership Society). 1999. Development of the 21st Century Workforce: Leadership, Commitment, and Action—The Crucial Next Steps.

Plack, M. M. 2003. Learning communication and interpersonal skills essential for physical therapy practice: A study of emergent clinicians. Columbia University Teachers College. Retrieved from http://proxy-remote.galib.uga.edu/login?url=http://search. ebscohost.com/login.aspx?direct=true&db=cin20&AN=20050 66014&site=eds-live Available from EBSCOhost cin20 database.

Preskill, H. and Torress, R. T. 1999. The role of evaluation inquiry in creating learning organizations. *Organizational Learning and the Learning Organization: Developments in Theory and Practice.* (M. Easterby-Smith, L. Araujo, and J. Burgoyne Eds.). London: Sage Publications.

Price, J. H., Akpanudo, S., Dake, J. A., and Telljohann, S. K. 2004. Continuing-education needs of public health educators: Their perspectives. *Journal of Public Health Management Practice,* 10(2), 156–163.

Probst, G. 1997. *Organizational Learning: The Competitive Advantage of the Future.* London: Prentice Hall.

Ranmuthugala, G., Plumb, J. J., Cunningham, F. C., Georgiou, A., Westbrook, J. I., and Braithwaite, J. 2011. How and why are communities of practice established in the healthcare sector? A systematic review of the literature. *BMC Health Services Research,* 11, 273. https://doi.org/10.1186/1472-6963-11-273

Ranson, S. L., Boothby, J., Mazmanian, P. E., and Alvanzo, A. 2007. Use of personal digital assistants (PDAs) in reflection on learning and practice. *Journal of Continuing Education in the Health Professions,* 27(4), 227–233. Retrieved from https://www.ncbi.nlm.nih.gov/pubmed/18085601

Ratmawati, D. and Herachwati, N. 2007. *Organizational Behavior.* Jakarta: Penerbit Universitas Terbuka.

Richardson, B. and Cooper, N. 2003. Developing a virtual interdisciplinary research community in higher education. *Journal of Interprofessional Care,* 17(2), 173–182.

Riegelman, R. and Persily, N. A. 2001. Health information systems and health communications: Narrowband and broadband technologies as core public health competencies. *American Journal of Public Health,* 91(8), 1179–1183.

Robbins, S. P. 2003. *Organizational Behavior* Jakarta: PT. Indeks Kelompok Gramedia.

Robbins, S. P. and Judge, T. A. 2013. *Organizational Behavior Edition 15.* New Jersey: Pearson Education.

Robbins, S. R. and Judge Timothy, A. 2017. *Organizational Behavior* (16th ed.). Essex: Pearson Education Limited.

Rosenfeld, L. S., Gooch, M., and Levine, O. H. 1953. *Report on Schools of Public Health in the United States Based on a Survey of Schools of Public Health in 1950*. Public Health Service U.S. DHEW Pub. No. 276. Washington, DC: U.S. Government Printing Office.

Russell, J., Greenhalgh, T., Boynton, P., and Rigby, M. 2004. Soft networks for bridging the gap between research and practice: Illuminative evaluation of CHAIN. *British Medical Journal (International Edition)*, 328(7449), 1174–1177. https://doi.org/10.1136/bmj.328.7449.1174

Salinsky, E. 2010. Governmental Public Health: An Overview of State and Local Public Health Agencies. National Health Policy Forum.

Sargeant, J., Curran, V., Jarvis-Selinger, S., Ferrier, S., Allen, M., Kirby, F., and Ho, K. 2004. Interactive on-line continuing medical education: Physicians' perceptions and experiences. *The Journal of Continuing Education in the Health Professions*, 24(4), 227–236. https://doi.org/10.1002/chp.1340240406

Sasser, J. n.d. Care Career. Retrieved from https://info.nhanow.com/blog/what-is-continuing-education-why-is-it-important-in-healthcare

Schein, E. H. 1985. *Organizational Culture and Leadership*. San Francisco: Jossey-Bass, Inc.

Schermerhorn, J. R. 2002. *Management*. (7th ed.). New York: John Wiley & Sons Inc.

Schemerhorn, J. R., Hunt, J. G. and Osborn, R. N. 2002. *Organizational Behavior*. New York, NY: John Wiley & Sons, Inc.

Schermerhorn, J. R., Hunt, J. G., Osborn, R. N., and Uhl-Bien, M. 2010. *Organizational Behavior* (11th ed.). New Jersey: John Wiley & Sons, Inc.

Schneider, M. and Schneider, H. S. 2017. *Introduction to Public Health*. Burlington, MA: Jones & Bartlett Learning.

Schweitzer, D. J. and Krassa, T. J. 2010. Deterrents to nurses' participation in continuing professional development: An integrative literature review. *The Journal of Continuing Education in Nursing*, 4141(1010), 441–447. https://doi.org/10.3928/00220124-20100601-05

Sedarmayanti. 2009. *Human Resources and Work Productivity*. Bandung: CV Mandar Maju.

Senge, P. M. 2006. *The Fifth Discipline: The Art and Practice of the Learning Organization*. New York: Doubleday/Currency.

Sheps, C. G. 1976. *Higher Education for Public Health: A Report of the Milbank Memorial Fund Commission*. New York: Prodist.

Smith, P. J. and Sadler-Smith, E. 2006. *Learning in Organizations: Complexities and Diversities*. London: Routledge.

Song, J. H., Chermack, T. J., and Kim, W. 2013. An analysis and synthesis of DLOQ-based learning organization research. *Advances in Developing Human Resources*, 15(2), 222–239.

Stein, D. S. and Imel, S. 2002. Adult learning in community: Themes and threads. *New Directions for Adult and Continuing Education*, 95, 93–97.

Strickland, S. P. 1972. *Politics, Science, and Dread Disease: A Short History of United States Medi cal Research Policy*. Cambridge, MA: Harvard University Press.

Supartha, W. G. and Sintaasih, U. K. 2017. *Introduction to Organizational Behavior*. Denpasar Timur: CV. Setia Bakti.

Taub, A., Allegrante, J. P., Barry, M. M., and Sakagami, K. 2009. Perspectives on terminology and conceptual and professional issues in health education and health promotion credentialing. *Part of a Special Section Entitled The Galway Consensus Conference: Achieving Excellence in Credentialing for Global Health Promotion*, 36(3), 439–450. https://doi.org/10.1177/1090198109333826

Terris, M. 1959. The changing face of public health. *American Journal of Public Health*, 49, 1119.

Thackeray, R., Neiger, B. L., and Roe, K. M. 2005. Certified health education specialists' participation in professional associations: Implications for marketing and membership. *American Journal of Health Education*, 36(6), 337.

Turnock, B. J. 2001. *Public Health: What It Is and How It Works*. Gaithersburg, MD: Aspen Publishers, Inc.

Ültanir, E. 2012. An epistemological glance at the constructivist approach: Constructivist learning in Dewey, Piaget, and Montessori. *International Journal of Instruction*, 5(2), 195–212.

University of Georgia, College of Public Health. 2013. College of Public Health Reaccreditation. Retrieved from https://publichealth.uga.edu/about/history-facts/

U.S. Centers for Disease Control and Prevention 2000. Public opinion about public health—United States, 1999. *Morbidity and Mortality Weekly Report* 49, 258–260.

U.S. DHHS (U.S. Department of Health and Human Services). 1988. *Bureau of Health Professions: Selected Summary Data on Fiscal Years 1980–1987 Awards*. Washington, DC: U.S. DHHS.

Wanto, H. S and Suryasaputra, R. 2012. The effect of organizational culture and organizational learning towards the competitive strategy and company performance (Case study of East Java SMEs in Indonesia: Food and beverage industry). *Information Management & Business Review*, 4(9), 467–476.

Watkins, K. and O'Neil, J. 2013. The dimension of the learning organization questionnaire (the DLOQ): A nontechnical manual. *Advances in Developing Human Resources*, 5(2), 133–147. https://doi.org/10.1177/1523422313475854

Watkins, K., Milton, J., and Kurz, D. 2009. Diagnosing the learning culture in public health agencies. Retrieved May 5, 2013, from http://hdl.voced.edu.au/10707/51261

Watkins, K. E. and Dirani, K. M. 2013. A meta-analysis of the dimensions of a learning organization questionnaire: Looking across cultures, ranks, and industries. *Advances in Developing Human Resources*, 15(2), 148–162. https://doi.org/10.1177/1523422313475991

Watkins, K. E. and Marsick, V. J. 1993. *Sculpting the Learning Organization: Lessons in the Art and Science of Systemic Change* (1st ed.). San Francisco: Jossey-Bass.

Wenger, E. 1998. Communities of practice: Learning, meaning, and identity. *Systems Thinker*, 9, 2–3.

Wenger, E., McDermott, R. A., and Snyder, W. 2002. *Cultivating Communities of Practice: A Guide to Managing Knowledge.* (E. Wenger, R. McDermott, and W. M. Snyder, Eds.). Boston: Harvard Business School Press.

Wild, E. L., Richmond, P. A., de Mero, L., and Smith, J. D. 2004. All kids count connections: A community of practice on integrating child health information systems. *Journal of Public Health Management & Practice*, 10, 61–65.

Wilner, D. M. 1973. *Introduction to Public Health* (6th ed.). New York: McMillan.

Winslow, C.-E.A. 1923. *The Evolution and Significance of the Modern Public Health Campaign*. New Haven, CT: Yale University Press; reprinted in 1984 by the Journal of Public Health Policy.

Winslow, C. E. A. 1953. *The Accreditation of North American Schools of Public Health*. New York: American Public Health Association.

Index

186 ■ Index